Teaching College Algebra

*Reversing the Effects of
Social Promotion*

Sherman N. Miller

Rowman & Littlefield Education
Lanham, Maryland • Toronto • Oxford
2005

Published in the United States of America
by Rowman & Littlefield Education
A Division of Rowman & Littlefield Publishers, Inc.
A wholly owned subsidiary of The Rowman & Littlefield Publishing Group, Inc.
4501 Forbes Boulevard, Suite 200, Lanham, Maryland 20706
www.rowmaneducation.com

PO Box 317
Oxford
OX2 9RU, UK

Copyright © 2005 by Sherman N. Miller

All rights reserved. No part of this publication may be reproduced, stored in a retrieval system, or transmitted in any form or by any means, electronic, mechanical, photocopying, recording, or otherwise, without the prior permission of the publisher.

British Library Cataloguing in Publication Information Available

Library of Congress Cataloging-in-Publication Data

Miller, Sherman N., 1942–
   Teaching college algebra : reversing the effects of social promotion / Sherman N. Miller.
      p. cm.
   Includes bibliographical references and index.
   ISBN 1-57886-242-6 (pbk. : alk. paper)
   1. Algebra—Study and teaching (Higher) 2. Mathematical readiness. 3. Mathematics—Social aspects. I. Title.

QA159.M55 2005
512'.0071'1—DC22
                                                                                    204065320

♾™ The paper used in this publication meets the minimum requirements of American National Standard for Information Sciences—Permanence of Paper for Printed Library Materials, ANSI/NISO Z39.48-1992.
Manufactured in the United States of America.

To my wife of forty-one years, **GWYNELLE W. MILLER**, who has been my editor of over 600 newspaper editorials, numerous letters to the editor, articles for magazines and trade publications, et al. I greatly appreciate her candid opinions in our discussions of various works.

In Memory of Dr. Edward H. Kerner

Dr. E. H. Kerner was an excellent teacher in the Department of Physics at the University of Delaware. He offered comments on this development and many other works over the thirty-years-plus close friendship with this writer.

Contents

List of Figures and Tables	vii
Introduction	1
CHAPTER 1 Developing a New Teaching Style	23
CHAPTER 2 Testing Initial Teaching Premises	31
CHAPTER 3 Testing Teaching Premises in Pre-calculus Course	45
CHAPTER 4 Preparing Nontraditional Students for College Level Mathematics	55
CHAPTER 5 First Semester Pre-calculus Under a New Paradigm	69
CHAPTER 6 Testing	169
References	175
Index	181
About the Author	187

LIST OF FIGURES AND TABLES

FIGURES

5.1	Five Groups of Three	71
5.2	Real Number Line, Sign Direction	73
5.3	Real Number Line	73
5.4	Finding Independent Axis Crossing	82
5.5	Coordinate System	91
5.6	Right Triangle	91
5.7	The Length of C = B - A	92
5.8	Distance Formula Derived	93
5.9	Applegate Produce Sales Bar Chart	97
5.10	Applegate Produce Sales Line Graph	97
5.11	R and W represent functions	99
5.12	Linear Form Plot	102
5.13	Second Degree Polynomial Plot	103
5.14	Radical Form Plot	104
5.15	Reciprocal Plot	105
5.16	Not A Function, Line Intersects Two Times	106
5.17	A Function, Line Intersects At One Point	107
5.18	Yearly Raise Projection	109
5.19	Definition of Increasing Function	110
5.20	Decreasing Function	111

5.21	Constant Equation	112
5.22	Relative Maximum	113
5.23	Relative Minimum	114
5.24	Absolute Minimum	114
5.25	Even Function	115
5.26	Odd Function	115
5.27	Neither Even nor Odd Function	116
5.28	Absolute Value	117
5.29	Vertical Graph Shift	117
5.30	Horizontal Graph Shift	118
5.31	Reflection Expression	119
5.32	Contraction Equation	120
5.33	Expansion Equation	121
5.34	Square	131
5.35	Rectangle	131
5.36	Circle	132
5.37	Triangle	133
5.38	Cube	133
5.39	Rectangular Solid	134
5.40	Half Sphere	134
5.41	Sphere	134
5.42	Circular Cylinder	135
5.43	Intercepts	136
5.44	Actual Graph Vs Estimation	137
5.45	Interception of Two Curves	138
5.46	Positive & Negative Coefficient Plot	144
5.47	Fourth Degree Polynomial Interceptions	145
5.48	Synthetic Division	148
5.49	Double Inequality	158
5.50	Salary	158
5.51	Credit Rating	158
5.52	Exponential Teaching Model Eight Week Accelerated Course	162

TABLES

5.1	Partial Times Tables	71
5.2	Standard vs. Solution Method	89
5.3	Student Generated Numbers	96

5.4	Stem and Leaf Method	96
5.5	Linear Form	101
5.6	Second Degree Polynomial	102
5.7	Radical Form	103
5.8	Basic Rational Function	105
5.9	Constant Equation	111
5.10	Contraction Equation	119
5.11	Graphing Aids for Functions	123

Introduction

It is enchanting to write off groups that we feel have beliefs or appearances different from our own. We overlook the great human potential these groups offer and merely vilify these people as societal outcasts. Yet, the real strength of the United States of America is our ability to find a diamond-in-the-rough and polish this stone to the level that its great value shines through for the world to see.

Let us use this mind-set to broach the idea that the African American community is an American uncut diamond in the rough. We first need to look back at the African American community's evolution into America's economic mainstream, where we ought initially focus our attention on some ingredients necessary for attaining upward mobility in the economic mainstream. A good starting place to find upward mobility ingredients is to examine allegories of our Founding Fathers because these leaders possessed a keen sense of what citizens needed for social advancement and maintaining it.

Today African Americans are struggling to overcome a legacy of slavery. Hence, one might expect this history to have the potential to taint one's psyche and cloud one's vision when viewing various upward mobility options. On the other hand, African American leaders became keenly aware of what is necessary to participate in the American dream. Black leaders knew that education is the conduit whereby one can rise from the ranks of America's have-nots to the esteem of mainstream America's haves.

Introduction

Thomas Jefferson made a high valuation of education in gaining and maintaining freedom of the people to Dupont de Nemours in April 1816 (Monticello, 2003). "Enlighten the people generally, and tyranny and oppressions of body and mind will vanish like evil spirits at the dawn of day."

In September 1818, Jefferson highlighted the importance of the public education of America's masses to Joseph C. Cabell. "A system of general education, which shall reach every description of our citizens from the richest to the poorest, as it was the earliest, so will it be the latest of all the public concerns in which I shall permit myself to take interest."

In 1822, Jefferson homed in on the importance of education in the pursuit of happiness to C. C. Blatchly. "I look to the diffusion of light and education as the resource to be relied on for ameliorating the condition, promoting the virtue, and advancing the happiness of man."

An attribute to Benjamin Franklin further corroborates the importance of education (Petrie, 2003). "A nation of well-informed men who have been taught to know and prize the rights which God has given them cannot be enslaved. It is in the region of ignorance that tyranny begins." We might argue that these allegories by our Founding Fathers underpinned the ideals of a master plan for African American upward mobility once slavery ended. But Blacks' hopes of advancing in the economic mainstream through utilizing America's public education system became an illusion of grandeur in 1896.

The 1896 U.S. Supreme Court ruling in the Plessy vs. Ferguson case put into law artificial barriers to stymie Black American upward mobility into Mainstream America. This action gave the United States of America the "separate but equal" doctrine, a catalyst that kindled the Age of Jim Crow. To look at this action solely from the triumph of white racists legitimating racial segregation in the national psyche would be to miss their real hidden agenda.

America's civil rights organizations from the beginning led Black America on a quest to knock down the evils of racial segregation. Everyone believed that if segregation were conquered, tomorrow would be a brighter day. In addition, Black American pride and dignity would naturally follow.

However, the black leadership fell prey to an old military trick in fighting wars. Good generals know that your goal is to make your enemy

Introduction

mass his or her forces on an illusion whilst you direct your strength at his or her weakest point. Racial integration evolved into an illusionary quest by Blacks whilst Mainstream America built new barriers around the Nation's prosperity.

A new paradigm on the prospects of having racial integration in the United States of America came with the unanimous 17 May 1954 U.S. Supreme Court ruling banning racial segregation in public schools (Dolmatch, 1982, p.101).

The 1998 World Book, in the article, "African American Journey: Brown v. Board of Education of Topeka," offers us a holistic look at the impact of this Supreme Court decision on changing the socioeconomic plight of minority group people (The National Center for Public Policy Research, 2003). "The Supreme Court's decision launched the legal movement to desegregate U.S. society. . . . The National Association for the Advancement of Colored People, guided by its chief lawyer, Thurgood Marshall, decided to use the Brown case and its companion cases to challenge the "separate but equal" principle. In the Brown case itself, Oliver Brown, an African American railroad worker in Topeka, Kansas, sued the Topeka Board of Education for not allowing Linda Brown, his daughter, to attend Summer Elementary School, in an all-white school near her home. The other cases involve similar suits by black parents from other parts of the country. Marshall attacked the "separate but equal" rule by arguing that segregation harms minority students by making them feel inferior and thus interfering with the ability to learn."

The World Book goes on, "In a unanimous decision, the court agreed with Marshall and declared that separate educational facilities could never be equal. Therefore, segregated schools violated the Fourteenth Amendment to the Constitution of the United States, which requires that all citizens be treated equally."

In the midst of the euphoria over black Americans being able to go to public schools, the black leadership paid little attention to the point the U.S. Supreme Court put the nation on a path to desegregate the U.S. society. Desegregation and racial integration became interchangeable concepts in the Black community; however, they carried very distinct differences in the economic mainstream (Miller, January 1999, p. 5).

Black America's blind pursuit of desegregation, mistaking it for racial integration, gave the anti-integration forces the ammunition to undermine the esteem of public education in America's psyche by making school-busing, the remedy necessary to desegregate public schools, into a highly charged word. White flight away from neighborhoods where busing would be the rule became an everyday occurrence in America. This white exodus left the black leadership merely watching the defrocking of the quality in and losing of mainstream support for public education without their fully comprehending the impact of this erosion in mainstream public support on the black community's evolution into the economic mainstream.

Thirty years after the Brown vs. Board of Education decision, the egalitarian dream of the black leadership did not reach the masses of black people. In 1987, there was a significant difference in literacy between black Americans and white Americans. Let us call on the literacy definition in the Young Adult Literacy Assessment advanced by Kirsch & Jungeblut as a frame of reference for our discussion. They offer the literacy definition, "Using printed and written information to function in society, to achieve one's goals, and to develop one's knowledge and potential" (Kirsch & Jungeblut, 1986).

What is disquieting for the black community is that literacy is the key issue in achieving equalitarian status in the U.S. economic mainstream, but black leaders focused their main efforts on black Americans going to a desegregated educational system whilst missing this upward mobility point. Richard L. Venezky, Carl F. Kaestle, and Andrew M. Sum highlight the level of deficient literacy in black America even including educated African Americans.

Venezky, et al, report on findings of the National Assessment of Educational Progress (NAEP) Young Adult Literacy Assessment for each of four major race/ethnic groups. They contend that "the race/ethnic patterns of differences in mean scores are identical for the NAEP reading and each of the three literacy test scales. White young adults achieve the highest mean scores on each of the reading and literacy scales, followed by other nonwhite Americans (Asians, and Native Americans), Hispanic Americans, and then black Americans. The absolute size of the gaps between the mean scores of white and black Americans exceed 50 points, or one full standard deviation. The mean scores of Hispanic young adults typically tend to fall halfway between those of white and black young adults and range between 0.5 and 0.7 standard deviations below those of white young adults" (Venezky, Kaestle, & Sum, 1987, pp 31–32).

Venezky, et al, continued, "A portion of the difference between the mean scores of White Americans and those of black and Hispanic Americans is attributable to differences in their formal educational attainment. Both black and Hispanic Americans were nearly twice as likely as white Americans to have terminated their schooling without receiving a high school diploma, and white Americans (40 percent) were nearly twice as likely as black (21 percent) and Hispanic Americans (21 percent) to have completed two or more years of college. While the score advantages of white young adults are due, in part, to their greater schooling attainment, large differences remain between white and black young adults even when controlling for years of schooling completed. . . . In general, Hispanic scores lag about 0.5 standard deviation units behind white scores for equal education attainment while black scores typically lag about 1.0 standard deviation behind white scores. Thus, the overall score differences between race/ethnic groups tend to be reduced by only a small amount when educational attainment is considered. The remarkable size of the difference is reflected in a comparison of the mean scores of black four-year college graduates with white high school graduates. . . . The latter group performs nearly as well even though they have had four years less schooling. Tabulations by parents' education show roughly the same differences, with children of white high school graduates outscoring children of Hispanic and black high school graduates by about 0.6 and 1.0 standard deviations, respectively. When all of the variables monitored by the survey are controlled— background, demographics, education, and literacy practices—only about one-third of the race/ethnicity differences disappear; the remaining differences are still quite large."

Clearly, the lessons of Jefferson and Franklin got lost in the euphoria over the legitimating of public school desegregation. Jefferson pointed out the need of education for the pursuit of happiness. *"What the masses of Black America found was the under education of Black children and their community snorting from a bout of purportedly too substandard literacy skills to fully participate in the bounty of the economic mainstream even when they get educated."*

Of course, there are always those people who manage to excel even in a deteriorating system. The "Black Talented Tenth" evolved under the tutelage of black leaders like Dr. W. E. B. Du Bois and showed the propensity to evolve even though Jim Crow became the order of the day. In 1905, Du Bois wrote, "Ignorance is a cure for nothing. Get the very best training possible and the doors of opportunity will open before you as they are flying before thousands of your fellows. On the other hand, every time a colored person neglects an opportunity, it makes it difficult for others of the race to get such an opportunity" (Katz, 1969, p. 364). There is little doubt here that the link between upward mobility in mainstream America and educational attainment was entrenched in the minds of the early black leadership.

On September 24, 1965, President L. B. Johnson's Executive Order 11246 enforced affirmative action for the first time. "The executive order requires government contractors to 'take affirmative action' toward prospective minority employees in all aspects of hiring and employment." Contractors must take specific measures to ensure equality in hiring and must document these efforts. On October 13, 1967, the order was amended to cover discrimination on the basis of gender.

Initiated in 1969 by President Richard Nixon, the "Philadelphia Order" was the most forceful plan thus far to guarantee fair hiring practices in construction jobs. Philadelphia was selected as the test case because, as Assistant Secretary of Labor Arthur Fletcher explained, "The craft unions and the construction industry are among the most egregious offenders against equal opportunity laws . . . openly hostile toward letting blacks into their closed circle." The order included definite "goals and timetables." As President Nixon asserted, "We would not impose quotas, but would require federal contractors to show "affirmative action" to meet the goals of increasing minority employment" (Brunner, 2000 – 2003).

Presidents Johnson and Nixon entrenched the idea of Affirmative Action into the psyche of the economic mainstream. These presidential actions and others that followed legitimated the "Black Talented Tenth." The doors of opportunity opened and the Black Talented Tenth rushed in fast as they knew that it would be foolhardy to miss this new quasi U.S. Reconstruction Era. The black middle class and upper class began to flourish, but this newfound black wealth did not reach the black masses (Miller, February 1999).

The advent of the legitimized Black Talented Tenth offered the racial segregationists a bonanza in their effort to hold the masses of black Americans socioeconomically deprived. It radically reduced the potential evolution of future black leaders with business acumen by vilifying educated blacks in the black community. Today, the black community often views their "Talented Tenth" with great disdain. Many everyday

Introduction

blacks believe their "Talented Tenth" sold out the black community to become corporate white clones. They view corporate black males as "Boy Toys" because they are mere toys for the mainstream business leaders to exploit for their own pleasure.

Norman Lockman, an African American Pulitzer Prize winning columnist now working with the *News Journal* newspaper in Wilmington, Delaware, highlighted in 1995 how this black leadership schism rendered local Wilmington blacks impotent. ". . . I [have] suggested," wrote Lockman, "that there were two kinds of black leaders, separate and unequal: community based leaders, and corporate and business leaders. Some way must be devised to get them working together" (Lockman, 1995).

He went on, "(African American) Mayor Jim Sills of Wilmington could do it. He has standing with both camps. Some callers said he won't try because of political considerations. Community leaders can help get out the votes. Corporate and business leaders tend to live outside of the city. Would Sills want to risk alienating community leaders by suggesting they are not as competent as the corporate and business types? In an election year? My telephone pundits didn't think so, but I think he could. Sills now relies on the community types for credibility and votes and the corporate and business types for organizational and financial support. Why not make that kind of power combo work on a broader scale for nonpolitical community interest?"

Hence, one senses the black community's embarrassment at seeing some purported self-appointed black leaders working hard to maintain segregationists' artificial barriers to black socioeconomic parity in the United States of America. That is to say, socioeconomic parity is the real goal and school desegregation is merely an element in this socioeconomic struggle. Lockman's comments suggest that the national black leadership pursued the wrong long-term objective when it focused solely on societal desegregation. Nevertheless, the Black Talented Tenth knew the real issue was black socioeconomic parity in the United States. Thus, we ought to look through these upwardly mobile blacks' eyes at modern actions that resulted in the unintended consequences of rekindling past segregationists' goals.

Gary Orfield, Mark Bachmeier, David R. James, and Tamela Eitle offer a regressive picture of public school integration in the late 1990s (Orfield, Bachmeier, James & Eitle, 1997). They highlight the role of the current U.S. Supreme Court in exacerbating racial and ethnic segregation in America.

Orfield, et al, report, "By far the most important changes in policy in the 1990s came from the Supreme Court. The appointment of Justice Clarence Thomas in 1991 consolidated a majority favoring cutting back civil rights remedies requiring court-ordered changes in racial patterns. In the 1991 Board of Education of Oklahoma City v. Dowell decision, the Supreme Court ruled that a school district that had complied with its court order for several years could be allowed to return to segregated neighborhood schools. In the 1992 Freeman v. Pitts decision, the Court made it easier to end student desegregation even when the other elements of a full desegregation order had never been accomplished. Finally, in its 1995 Jenkins decision, the Court's majority ruled that the court-ordered programs designed to make segregated schools more equal educationally and to increase the attractiveness of the schools to accomplish desegregation through voluntary choices were temporary and did not have to work before they could be discontinued."

They concluded, "In other words, desegregation was redefined from the goal of ending schools defined by race to a temporary and limited process that created no lasting rights and need not overcome the inequalities growing out of a segregated history."

In 1969, President Richard M. Nixon incorporated a system of "goals and timetables" to evaluate federal construction companies according to Affirmative Action criteria. However, in 1999, the *Wall Street Journal* reports, "Moving to conform with a 1995 Supreme Court ruling, the Clinton administration is scaling back its affirmative action program for federal transportation projects. The Transportation Department announced that it is dropping requirements that all contracts for federally assisted highway, airport and transit projects have a target for participation by disadvantaged businesses, largely those owned by women and minorities. Instead, states, transit authorities and other agencies that allocate federal transportation aid will be expected to promote affirmative action primarily through broad, race-neutral programs of technical and other assistance" (Wall Street Journal, 1999).

During his presidential tenure, former President Bill Clinton allowed affirmative action to move to the back burner where blacks, women, et al, needed to accept that this safety net is now a relic of a by gone era. Some liberals might argue that there is still hope to regain the affirmation action of yesterday, but these people learned nothing from the end of the Reconstruction Era when the national psyche shifted to accepting destructive race relations as the norm. It becomes easy for many of these affirmative action proponents to merely debase Republican Conservatives in the media as a throwback to yesterday's racists rather than to ask if these Republican leaders have a viable alternative method of achieving minority socioeconomic parity in America.

The battle over affirmative action is boiling down to answering the question, "When do yesteryear's disadvantaged group members no longer need remedial help to become competitive in the United States of America's economic mainstream?" Heretofore we discussed the impact of education as the conduit to upward mobility in the United States of America, and now we are suggesting that the mainstream is no longer tolerant of substandard performance by any group. This line of thought gained legitimacy in July 1995 when the University of California Board of Regents came out against preferential treatment (affirmative action) in admitting students to the university. Their action was a catalyst for the state of California passing Proposition 209 that outlawed preferential treatment based on race, sex, color, ethnicity, or national origin in public employment, public education, and public contracting.

California Regents chair Tirso del Junco, in a November 6, 1996, statement, clearly shows the board relished the legitimacy of their position with the passage of Proposition 209. "With the passage of Proposition 209, the citizens of California have affirmed the Board of Regents' July 1995 decision to end the use of race, ethnicity, and gender in the University of California's admissions, employment, and contracting practices," stated del Junco.

However, del Junco then faced up to the Herculean task of making such a radical change in direction work. "This challenge will require the creativity and ingenuity of everyone in the University community – regents, administration, faculty, students, and staff" (del Junco, 1996).

Without going through all of the University of California's policy, "Policy Ensuring Equal Treatment Admissions (SP-1)," two of the nine sections merit serious pondering for their potential to shape national policy.

> Section 2. Effective January 1, 1997, the University of California shall not use race, religion, sex, color, ethnicity, or national

origin as a criterion for admission to the university or to any program of study.

Section 9. Believing California's diversity to be an asset, we adopt this statement: Because individual members of all of California's diverse races have the intelligence and capacity to succeed at the University of California, this policy will achieve a UC population that reflects this state's diversity through the preparation and empowerment of all students in this state to succeed rather than through a system of artificial preferences.

The above statements have the potential to produce excellent media sound bites that provoke the ardor of Mother's Day, apple pie, and the flag. Although the statements are nightmares for the proponents of affirmative action, Section 9 does pull the rug from under those intellectual racists espousing the intellectual inferiority of African Americans. The Regents left no doubt on their belief in the intellectual capacity of all of California's multifaceted racial and ethnic population saying, " . . . Individual members of all of California's diverse races have the intelligence and capacity to succeed at the University of California."

If we look at SP-1 from an affirmative action point of view, we are limiting access to education at some of the nation's finest institutions of higher learning for blacks, women, and other minorities. The affirmative action forces argue that yesteryear's disadvantaged groups are not ready to have their affirmative action security blanket removed because poor quality education is often the norm in many of their neighborhoods. Substandard education now makes these disadvantaged minorities academically unprepared for mainstream rigor.

The anti-affirmative action psyche took hold in Texas where it radically changed admission policies at Texas A & M University by encumbering their ability to recruit black and Hispanic students. The Hopwood decision underpinned a shift to race and ethnic neutral admissions policies at Texas A & M. Gary Engelgau highlights the initial Texas decision, writing in ON Target—College Board On-line.

Engelgau writes, "The initial decision in the Hopwood case was made by the Texas Supreme Court in 1995. The court ruled that the University of Texas Law School's affirmative action plan was unconstitutional because separate admission committees were used to review minority and non-minority applications for admission. " (Engelgau, 1998–1999).

The impact of the anti-affirmative action mind-set portends a bleak future for blacks and Hispanics when it comes to participating in the bounty of the American economic mainstream. Everyone agrees that education is the key to upward mobility in the United States and to close that door is a death-knell to any group.

Brian Fishman offers the aftermath of California Proposition 209 in the *Daily Bruin*. He writes, "While protests, arrests, and vicious debate have surrounded the repeal of affirmative action, diversity at UCLA Law School has continued to decline.

"This year, the number of African American applicants dropped 55 percent from 1994 levels. The number of Latino applicants has also fallen by 44 percent since 1994, the year before the decision to discontinue affirmative action.

"During the same period, the number of African Americans actually admitted to the law school dropped from 26 percent to 8 percent while the number of Latino admits fell from 17 to 12 percent" (Fishman, 1998).

Fishman's report is very sobering to the black and Latino communities because it truly quantifies the loss of immediate upward mobility resulting from the demise of affirmative action. The symbolism of the fallout from Proposition 209 is for black and Latino students to realize that the welcome mat at California public universities now says "unwelcome."

With affirmative action forces being crushed in California and Texas, surely some political leaders would capitalize on this momentum at the national level. U.S. Representative Frank Riggs (R-CA) stepped forth to seize this opportunity. Priti Kataria writes, "The amendment, proposed by Rep. Frank Riggs of California, would prohibit any public university from granting preferential treatment to any person based on their race, sex, ethnicity, or national origin. However, preferences given to anyone affiliated with a Native American tribe is still allowed" (Kataria, 1998, pp1–2).

Riggs mimicked Proposition 209. But he made the fatal mistake of failing to realize that the American people had data on which to judge the impact of this new proposed law, so his ability to exploit the ardor surrounding equating affirmative action to preferential treatment vanished.

Congresswoman Maxine Waters (D-CA) provided the proof of the deleterious impact of Proposition 209 in Riggs' home state of California. She stated, "Since the University of California System eliminated affirmative action, admissions for Black and Latino students have plummeted. Black undergraduate admission dropped 66 percent at UC Berkeley, 43 percent at UCLA, 46 percent at UC San Diego, and 36 percent at UC Davis. Latino undergraduate admissions dropped 40 percent at UC Berkeley, 33 percent at UCLA, 20 percent at UC San Diego, and 31 percent at UC Davis."

The Riggs Amendment to the Higher Education Act failed, by a vote of 249 votes against it to 171 votes for it. This Riggs failure suggests that the national leadership cannot afford to succumb to yesterday's policies that limited opportunity based on race and gender because successful participation in the global marketplace makes these efforts national socioeconomic suicide. Texas is working to get around the constraints of the Hopwood ruling as there is a radical shift occurring in the racial and ethnic make up of its work force.

A 1998 Texas Higher Education Status report offers plenty of food for thought on where the population is going in Texas in the near future. "Blacks and Hispanics made up approximately 11 percent and 22 percent, respectively, of the state's labor force in 1990. By 2030, blacks are expected to account for about 9 percent, but Hispanics are projected to represent nearly 46 percent of the work force in Texas, according to the Texas State Data Center at Texas A & M University. " (Higher Education in Texas, 1998, pp. 24 – 25).

The above figures offer Texas lawmakers little option but to find ways of opening opportunity to minority people or simply become noncompetitive in the global market from a labor pool that is too substandard. The Texas leadership is acting to nullify the potency of the Hopwood ruling.

The higher education status report also revealed, "The 75th Texas Legislature also took strong steps to ensure diversity in Texas higher education. House Bill 588 requires public universities in the state to automatically admit high school graduates who rank in the top 10 percent of their high school class."

It continued, "Universities also have the option of increasing automatic admissions to include students who rank in the top 25 percent of their high school class. After admitting students under those provisions, institutions must consider all, or a combination of 18 other factors when admitting other students. Examples of those criteria include the student's academic record, the student's socioeconomic background, the performance of the high school the student attended, the financial status of the student's school district, the student's personal responsibilities while attending high school, the student's performance on standardized tests, and the student's extracurricular activities."

Texas efforts prove the adage, "Where there is a will there is a way." The University of California saw the light. Syndicated columnist Bonnie Erbe writes, "In desperation, the UC system is now trying to adopt a measure to guarantee the top 4 percent of the graduating high school students a spot in its freshman class" (Erbe, 1999). Surely, the University of California could not bury their heads in the sand after the disquieting comments made by U.S. Representative Maxine Waters on the drop in their minority student enrollment in the wake of the passage of Proposition 209.

The affirmative action taint on college admissions is now spawning a redefinition of the concept of diversity. In the past it was a euphemism for black and white relations. The University of Delaware is now using a definition advanced by their director of admissions Larry Griffith.

Allison Taylor reports on an interview with Griffith, "It's difference in values and life-style that diversify a campus. 'It's not a student's racial background that's really going to mean anything. It's what your cultural and ethnic heritage is. That's what really [shows] perspective and, experience,' he said" (Taylor, 1999).

What Griffith is finally bringing to light is the fact that blacks include people from many nations in Africa, the Caribbean, and a host of other nations throughout the world. These people have the same skin color but their culture can be radically different. A similar scenario is true for people of European, Hispanic, and Asian decent.

The University of Delaware's efforts to diversify its student body whilst subordinating race suggests that a race neutral psyche for college admission is the dawning paradigm for the new millennium. Clearly, the efforts of Texas and California to devise methods to incorporate their diverse citizens into the educational mainstream are being watched by the nation. Nevertheless, we must ask ourselves what is working and what do we need to do to be able to avoid the advent of balkanization as the United States of America grows tanner.

In the midst of all of the haranguing over access to the economic mainstream through a quality education, we have not looked at the preparation needed to teach in this new educational arena. Chills run down your spine when you read a comment such as that reported by Anjetta McQueen of the Associated Press. "Washington — 'Four out of five U.S. teachers say they are not ready to teach in today's classrooms. And more than a third say they either don't have degrees in the subjects they teach or didn't spend enough time training in them,' says a survey released Thursday by the Education Department" (McQueen, 1999).

There are efforts underway to address this crisis in the classroom. McQueen continued, ". . . Last week, [President Bill] Clinton proposed spending more than $245 million to hire new teachers, train teachers for impoverished school districts and recruit teachers for areas heavily populated by American Indians. Congressional Republicans started hearings on teacher quality this week and will produce proposals by spring."

Although we may desire not to speak to the impact of incarceration on the upward mobility of minority people when it comes to education, we cannot bury our heads in the sand on this issue. Dr. Manning Marable makes it very uncomfortable for us to even flirt with ignoring the relationship between incarceration and education on the socioeconomic plight of America's minority communities.

"Today there are more than 1.8 million people who are incarcerated in federal facilities, state prisons and jails across the United States," writes Marable. "We now imprison more people than any other country on earth. Since 1980, about one thousand new prisons and jails have been built in the United States, at the cost of tens of billions of dollars. This elaborate network of coercive institutions is essentially a means of social control for black, Latino and low-income people. It is no accident, for example, that one half of all inmates are black; that one out of every fourteen black adult males, as you are reading this article, are currently in jail or prison; or that one out of every three African-American males in their twenties are today either in prison or jail, or probation, parole, or awaiting trial" (Marable, 1999).

Marable took away any doubt over the impact of this prison-building boom on upward mobility in the minority community. He continued, "A recent study produced by the Correctional Association of New York and the Washington, D.C.-based Justice Policy Institute, illustrates that in New York State hundreds of millions of dollars have been stolen from the budgets of public universities and reallocated to prison construction. The report states, 'Since fiscal year 1988, New York's public universities have seen their operating budgets plummet by 29 percent while funding for prisons has increased by 76 percent. In actual dollars, there has nearly been an equal trade-off, with the Department of

Correctional Services receiving a $761 million during that time while state funding for New York's city and state university system has declined by $615 million.'"

Kataria pointed out the hidden agenda against people with drug convictions that U.S. Representative Frank Riggs of California had in his defeated amendment to the Higher Education bill (Kataria, p. 1). This defeat was of great significance in the minority community when you consider the revelation in a 1997 government release of crime statistics. "Washington (AP) —Drug offenders accounted for nearly a third of the 872,200 felony convictions in state courts in 1994. " (Government releases latest crime statistics, 1997). Clearly, you can see if the Riggs amendment had passed it would exacerbate hopelessness in the black communities across the United States of America.

When you look at the background of those people going to prison, you find education is a missing element in these prisoners' lives. In a report emanating from California entitled, "Seeking Justice," we find in the section "Who Goes To Prison" a very disturbing commentary on prisoner educational levels. "It is estimated that 50 to 75 percent of all state prison inmates are unable to read. Only one-third of the prisoners nationwide have completed high school. By contrast, of the general population, 85 percent of all men ages 20 to 29 have high school diplomas" (The Edna McConnell Clark Foundation, 1997, p. 13).

If we believe that this absence of education leads to an entrapment in poverty, we see that case summed up well in comments by Pierre (Pete) Du Pont IV writing in the *Journal Gazette*. Du Pont is a conservative Republican (former U.S. Representative and former governor of the State of Delaware). He writes, ". . . Only two things have demonstrated they can break the hammerlock of welfare as a way of life for the poor—jobs and education. Numerous studies have repeatedly shown that young women who stay in school and graduate are far less likely to need state welfare for support.

"With high school skills, a worker has the chance to compete for real jobs— ones that offer training, advancement and a future.

"A high school education is the single crossover element that can both work to eliminate the increasing number of teen mothers, and help convert existing teen welfare mothers into workers able to compete for jobs in the real world" (Du Pont, 1994, p. 17).

Introduction

We have argued that during America's legalized racial segregation epoch our leadership built an infrastructure that closed quality education to America's nonwhite population. This system flourished whilst the United States of America was solely a Eurocentric population; hence, there were no economic or national security reasons to educate nonwhite Americans.

The pervasive racial segregationist's mind-set started to crumble in the decade of the 1960s with the passage of the Civil Rights and Voting Rights acts coupled with the U.S. Supreme Court's knocking down of the miscegenation laws, thereby legitimating nonwhites in America. The decade of the 1970s gave us a paradigm shift away from Eurocentric immigration as the norm because European immigrants plummeted to 18 percent during this period. This incarnated the tanning of America and it repositioned upward the need for nonwhites in the economic mainstream. Furthermore, in the midst of the present global "war on terrorism," the United States of American cannot afford to allow its minority group peoples to remain disenfranchised because the al Qaeda international terrorist organization may find these abandoned neighborhoods ideal havens for their sleeper cell operations.

America's corporate community must have an educated work force to remain competitive in the global marketplace. The military-industrial complex needs trained soldiers, scientists, and engineers for the security of the nation. Coupling the military and corporate needs for educated people and realizing that the pool from which these prospective workers and soldiers are coming is growing more and more nonwhite, then it is of paramount importance to educate America's nonwhite population.

Teachers are reporting that they are unprepared to teach because of deficiencies in their backgrounds. We are shifting to a race neutral educational system and economic mainstream. There are efforts underway to raise educational standards, such as the "No Child Left Behind Act of 2002" (Wright, p. 15).

In this current educational debate, we hear a great deal about the need to improve mathematics scores for children but do not see a great deal of debate over how to teach it. We hear even less about how to improve mathematics scores for people who passed through the school system and are now prisoners of ignorance and parents of children in today's schools.

Introduction

One way to help to improve public school education in the present national effort is to give poorly educated parents the necessary tools to help their own children. Hence, this writer wishes to share some teaching techniques of college algebra that evolved from teaching mathematics to adults students between the ages of 18 and 50.

We recognize that black Americans have had serious problems receiving a quality equation from many U.S. public schools in the last millennium. But we must also look at the pervasiveness of this poor quality education taint on the nation's public school education in general. What is the direction for the new millennium?

The Affirmative Action question of using race as a basis for college admission was solved in the June 2003 U.S. Supreme Court ruling in the University of Michigan case Barbara Grutter vs. Lee Bollinger, et al. It can be used as a factor to promote diversity. This is especially important with the browning of the United States population and the ability of American industry to remain competitive in today's and tomorrow's global business society.

Today's universities recognize that their educational role has changed with our enchantment with globalization. This enchantment has overlooked the fact that the United States of America now finds itself in the midst of a global economic war where tomorrow's prosperity is the bounty. The media regularly reports on thousands of jobs being lost, but little is being said about what the average American can do to hold onto her or his job tomorrow. However, the key to victory in this global economic war may be how well we develop all of the human capital of this nation.

In yesteryears, minority peoples were considered insignificant contributors to the nation's mental resource potential. Some have argued that some minorities are inherently inferior, and have attempted to make the case that minority peoples were put into this world solely to do the menial tasks. Hence, racial segregation laws were excellent barriers to upward mobility for nonwhites moving up into the economic mainstream because without a good education minority people were noncompetitive in mainstream America.

In 1954, the African American leadership marveled at the advent of the era of public school racial desegregation with the 1954 Supreme Court ruling in Brown vs. Topeka Board of Education where the infamous "separate but equal" doctrine was voted down. Students' busing became the desegregation remedy for America's public schools. But busing kindled the ire of white America and a white exodus from impacted neighborhoods left the black leadership grappling with the advent of new socioeconomic barriers to the use of African American talent in the economic mainstream.

Inner-city America became a wasteland overrun by crime and drugs. The felony conviction became a rite of passage for many inner-city youth. With inner-city children being vilified, there was no need to educate these people. A way to get the troublemakers through the public school system became the exploitation of "social promotion." However, a social promotion system suggested that the human resource potential the nation needed for today's global economic competitiveness did not exist in inner-city America. Social promotion's long-term deleterious impact on killing tomorrow's U.S. competitiveness is felt in the 1997 comment of the American Federation of Teachers: "Social promotion is particularly insidious, not only because the problems of failing students are ignored, but because it sends a message to every student that effort and achievement hardly matter.... If achievement is irrelevant to student progress, then teachers' ability to demand that all students meet high standards is seriously eroded" (Social Promotion, 2001).

It also should be pointed out that the United States does not possess a monopoly on vilifying people from crime-ridden neighborhoods. In the United Kingdom, the work of Les Johnston, Robert MacDonald, Paul Mason, Louise Ridley, and Colin Webster highlighted in *Findings* reveals, "Young people from Willowdene [Britain] faced problems and stigma as a consequence of its negative labelling (e.g. failing to get job interviews because of their postal address, criminal victimisation, low levels of recreational and service provision)" (Johnston, MacDonald, Mason, Ridley & Webster, 2000).

A gauge of how far the inner-city public school children have deteriorated educationally is seen in a comment reported by Debra Moffitt in the Wilmington, Delaware, the *News Journal*. "...The average [Wilmington] city high school student had a 'D' average last year and fewer than one in five seniors had plans for college," wrote Moffitt (Moffitt, 1999, p. B1).

We need also to have an appreciation of the Herculean task before us in offering teaching methods to improve the educational lot of nonwhite America in the twenty first century. Patricia A. Wasley shares the white teachers' mind-set in their having to teach nonwhite students. She writes in a book chapter entitled, "Responsible Accountability and Teacher Learning" that "Although most teachers worked hard to ensure that their students were successful, many grew to believe that poor children and children of color do not achieve as well as their more privileged White counterparts because they saw these differences play out in their own classrooms. It is also true that teachers teach the way they were taught themselves. . . . As students, many teachers were taught by teachers who did not believe that all children could learn. The bell curve, then, became a kind of self-fulfilling prophecy influencing teachers around the country for decades" (Wasley, pp. 136-137).

Subjects such as science and mathematics were areas to be shunned by inner-city students. Mediocrity became the norm for many people who were just trying to get out of high school with as little as possible academic effort. Consequently, many of today's inner-city adults who were pushed through the ill-famed public schools lack basic literacy. However, the natural ability is there and it only needs refining. Delaware State University developed a program entitled, "Project Success Wilmington Campus" (Miller, 2003), to open college opportunity for inner-city Wilmington nontraditional students.

A holistic look at the comments above suggests that people coming from inner-city public schools or metropolitan school districts that cater to inner-city children may need a great deal of support to become successful college students. Yet, this support is not always easy to get, for these challenged students need a real mentor program. "Real mentorship programs for minority students . . . are very scarce in predominantly white or private universities, and even in community colleges," says Sonia Delgado-Tall, an English professor at Roosevelt University. "Clubs, black student unions, cultural associations, student activities offices or multicultural programs cannot be substitutes for academic mentorship, where minority students could be paired with instructors . . . with whom they may feel [more] cultural kinship" (Forte, 2002, p. 5).

What is suggested here is that the role of the college professor will need to include mentoring to inner-city traditional and nontraditional students and first generation college students while these students are trying to acculturate the mainstream psyche. When it comes to mathematics, the professor-mentor will need to understand how the student learns.

David Tall of the Mathematics Education Research Center at the University of Warwick reports, "There is a broad spectrum of student thinking styles, partly genetic, partly influenced by social experience and teaching, which predispose students to different kinds of learning techniques. . . . It may be hypothesized that 'natural' learners build from their experience and try to make sense of the mathematics through their current knowledge, 'formal' learners are willing to take the mathematics and its rules as a game to be played, to make sense of it with itself" (Tall, 1997, p. 1). Less-chanced students who happen to lean towards natural learners are struggling because they may not have a good background from whence to draw.

Tall closes his paper indicating that no one learning style will fit all. "The evidence here shows that a single method will not work with all students. There is a role for the sensitive teacher aware of the needs of the student" (Tall, 1997, p. 19).

Several factors must be considered in getting students to perform better. The Rochester Institute of Technology (RIT) offers plenty of food for thought on the scope of this problem. "According to the TIMSS (Third International Mathematics and Science Study) in the early 1990's, American math education covers more concepts in its curriculum than are covered in other countries, does it more superficially, and emphasizes math skill over in-depth understanding" (Learning Mathematics, p. 2).

RIT continues, "It is no wonder that there are students who are unprepared to handle college level math courses, even those with four years of high school math and a 'B' average! These students' difficulties involve: memorizing math facts and formulas, not effectively using the math textbook, solving routine math problems and being tested more on skills versus understanding."

But the present scope of the education crisis with students' comprehension of mathematics is seen in a June 2003 *New York Times* article by Michael Winerip entitled, "A 70 Percent Failure Rate." ". . . The state's [New York] Council of School Superintendents estimates that 70 percent of those who took the state math test failed ..." (Winerip, 2003, section B, p. 9).

This present work was developed to help to improve the pass/failure rate of students majoring in business management who were required to take 100 level mathematics at an Eastern university. Its goal was "to develop teaching techniques whereby 80 percent of School of Management majors will pass the 100 level required mathematics courses (college algebra and finite mathematics) and 60 percent of the passers obtain a grade of 'C' or better with a target date the end of 2008." The focus of this effort is on teaching techniques for college algebra because it is the prerequisite course for finite mathematics, business calculus, and statistics that are required to obtain a degree in management. In a report entitled, "Student Success in College-level, Courses Assessment Report 2000," Northern Virginia Community College (NVCC) offers some guidance on setting a pass/fail goal. NVCC experience suggests that the above objective is achievable as they report, ". . .Of the students who had enrolled in a college level math class after completing a developmental math class, 47% successfully completed the course with a 'C' or better" (Student Success, 2000).

However, there may be some college and university professors recalcitrant towards altering their teaching style to encourage students' academic success, but these professors may find themselves victims of modern technology. Tenure may not prevent aggressive upstart Internet companies from using state laws to expose a teacher's pass/failure record to the world. University and college professors can no longer live a cloistered life where they feel that they are not accountable to the general public over their grading history. "Pick-A-Prof" is advertising that they have a half million students belonging to their Web site. They give histories on professors' grading patterns, hence allowing students to shun professors with poor grading histories. Therefore, college professors may be forced into paying a great deal of attention to pedagogy in the near future or fall victims to the wrath of an Internet defrocking, (Pick-A-Prof, 2004).

Chapter 1

Developing a New Teaching Style

It is amazing how strokes of fate over time have led to significant research developments. We may be looking for one thing only to find another because we assessed our observations beyond the sterile boundaries in which we initially believed to hold our discovery. Hence, it is an excellent idea to also recount events in our lives to garner the great learning they offer us today.

Teaching Through Passion

I taught freshman level mathematics at Delaware State University (formerly Delaware State College) whilst attending the University of Delaware to work on a graduate degree in physics. At the University of Delaware, I carried the added burden of being the first African American to major in physics.

The Delaware State mathematics course had roughly thirty students. Roughly half of the students were black and the other half white. Self-imposed intra-classroom racial segregation was the rule: all of the blacks sat on one side of the room whilst all of the whites sat on the other. My initial goal was to assess where the students' preparedness was and to set my lectures to the mean of the class. Hence, I gave a test that was designed to yield the class preparedness, expecting a mean test score between 40 and 60 for students who truly belonged in this course. I planned to encourage students getting test scores in the 80s and above to take a higher-level mathematics course.

When I handed back the tests during the next class, I heard grumbling in the rear of the classroom, something like, "I'm not taking this from no teacher!" Before I could catch my breath a big African American student stood up in the aisle next to his desk making demands. He had a haunting appearance and I sensed that he had gotten through high school intimidating teachers. It was clear that if I failed to get this student in line it would be a long semester.

This student made the fatal mistake of assuming that I was either a scared nerd or I had grown up in a suburban black neighborhood with no knowledge of how to handle myself on the streets. Having spent my early childhood growing up between two public housing projects, I had no problem returning to the 'hood on any given day, so I understood the term, "Wolf Ticket." He was selling wolf tickets (threats and bluffs) in hopes that I would acquiesce. Caving in is a sign of weakness in the 'hood, so I asked him to step outside so that we could resolve the issue. He quickly sat down.

I instructed the class that I would rule with an iron rod and no nonsense would ever be permitted in my classroom. But a second student had to see if I meant what I said. I asked him to go to the blackboard to do a problem and he went into a ghetto-style promenade where he showed off for his *soul* brothers and sisters (black companions). This chap worked out the problem incorrectly and turned around awaiting the approval of mates for his display of tomfoolery.

I took his work apart line-by-line showing how poorly his effort was. His demeanor changed from that of a show-off to a person willing to pay attention to what I had to say. I made it clear to him that the next time I called him to the blackboard I expected no procrastination. I did not have any other significant discipline problems throughout that semester. The word spread that I was a tough person and you had better not consider any foolishness in my classroom.

This experience made it clear that teachers must have respect. If I were to teach people from the 'hood and see that they excelled, I must stand tall in their view. I also showed the students that I cared about their success. I demanded performance; mediocrity was not an option.

Nevertheless, it is worth realizing that the threatening actions from the students above are not unique to the 1960s nor solely a United States of America problem. The U.S. Department of Justice reported in a report on school crime in 1991, ". . . Fifteen percent of the students said their school had gangs, and 16 percent claimed that a student had attacked or threatened a teacher at their school" (School Crime, 1991). This is disquieting when you consider that students are bringing all sorts of weapons to school in this new millennium.

BBC News reports, " The survey of local authority schools in Scotland found 743 incidents of violence against teachers in 1997-98, with the qualification that this figure is likely to be an underestimate because of under-reporting and incomplete returns from some schools" (Education Teachers Face Violence, 1999).

One might see my aggressive action with the two students above as a bit draconian, but failure to grab control of this problem very quickly results in national actions of desperation such as "zero tolerance" that one reads about in *Building Blocks for Youths*. "While students are reporting school crime at the same level as in the 1970s, the number of youth suspensions has nearly doubled from 3.7 percent of students in 1974 (1.7 million students suspended) to 6.8 percent of students in 1998 (3.2 million students suspended). African American students are suspended at roughly 2.3 times the rate of white students nationally" (Fact Sheet, 2003).

In 1996, the Florida legislators worked on General Bill H301: Teachers' Protection Act of 1996 where they were attacking the violence against teachers. This bill included a provision which "require[s] the school board to adopt rules for expulsion for at least one year of a student convicted of a forcible felony or violence against an educator" (Teachers' Protection Act, 1996). Hence, for people working solely in adult education, we must be prepared to address this violent mind-set in potential students that may follow them to adult settings.

Evolution of a Teaching Tenor

In the wake of the discipline problem I revealed above in my first teaching assignment at Delaware State University, I adopted a teaching style that fits the description of what Indiana University calls "*authoritative*." "The authoritative teacher places limits and controls on the students but simultaneously encourages independence. This teacher often explains the reasons behind the rules and decisions. If a student is disruptive, the teacher offers a polite, but firm, reprimand. This teacher sometimes metes out discipline, but only after careful consideration of the circumstances" (Classroom Management, 1996).

Indiana University offers three other styles: indifferent, laissez-faire, and authoritarian. With adults who may be raised in the 'hood, being overbearing or allowing the students to do whatever he or she wishes at the moment offers too great an opportunity for nothing productive to occur. If you are overbearing you ought not be surprised if these students rebel. If you allow them to set the learning agenda, they will only learn what they like and not necessarily what they need. Hence, a strong and compassionate teacher will offer the students the structure and the direction needed to ensure that they develop the skills necessary to be successful in mainstream American and global job markets.

Understanding Is Very Addictive

After my first semester of graduate studies at the University of Delaware, the university offered me a National Science Foundation Developmental Fellowship. This meant that I no longer taught mathematics at Delaware State University. The University of Delaware expected me to prepare myself to do physics research. However, I now had time to offer assistance to other black students who needed help at the University of Delaware. In addition to blacks who earned regular admission, the University of Delaware had a program called "College Try" where they recruited a number of black undergraduate students with substandard backgrounds to quickly grow their black population.

It was difficult to distinguish between College Try blacks and those not associated with that program. Therefore, blacks attending by virtue of their merit generally had to fight the stigma associated with this effort to rapidly expand the black population. It should be noted that there were some high caliber blacks on campus, many of whom have gone on to become successful citizens in the economic mainstream. Therefore, I will make some general comments on the two generic groups of black students. There were first people from the inner city who possessed hidden ability but lacked either disciplined work habits or confidence in themselves academically. Many of the inner-city blacks might give up under the academic strain; especially if they believed that no one expected them to succeed in the first place. Conversely, there were the Blacks, many of whom were from the rural communities, who possessed excellent talent and had excellent work habits. Rural blacks were driven to work hard and not give up easily because they knew the alternative to success in college meant a return to the hard and dirty farm work they were trying to escape.

Two undergraduate University of Delaware black students came to me a couple of days before their final examination, worried that they would fail college mathematics. One was a male from the small city of New Castle, Delaware, and the other a female from Delaware's largest city, Wilmington. The male said he was presently failing the course. They asked if there was anything I could do to help them pass the final examination.

Having taught a comparable course at Delaware State University the previous year, the course information was fresh in my mind. I felt compelled to take on this challenge.

We chatted awhile, giving me the opportunity to assess the task. I concluded that both students had merely listened to their professor lecture with no idea of what went on in the classroom. It was also obvious that they possessed a minor reading problem and their arithmetic background was questionable. Nevertheless, with time being our worst enemy, a crash program was our only option.

I first reviewed addition, subtraction, multiplication, division, and decimals for whole numbers and fractions. This offered a foundation on which to teach the higher-level mathematics. It also gave the students their first sense of accomplishment. But I did not give these students an assessment of the amount of time I really thought it would take to accomplish our task because I feared that they may not have wanted to see it through.

Once we finished the basic arithmetic, we proceeded to go through the book. We learned how to read definitions and understand the examples. Since time was short, we could not stay on any area too long. Nevertheless, we did not move forward until they understood the principal concepts in the chapter.

The hours passed and fatigue set in. We continued to move no matter how tired we all became because we had to finish that textbook in twenty-four hours. After about twelve hours, I did not have to tell them how long it was going to take because everyone was numb.

As the sun came up, we found that we were well on our way to completing the book. We worked on and it took roughly twenty-four hours to complete this text. There was insufficient time to test how well they had learned the material, so the students had to make this discovery taking their final examination.

According to the male student, several sections took the same final examination at the same time. Which meant a few hundred students would be compared to one another. In a 1999 chat he said, "I got the second highest score of all of the students taking the examination. My professor, who was a graduate student, said, 'Don't be surprised by your grade but you should have gotten an 'A' because I knew you did the examination, but I had to give you a 'B'.'"

This former student went on to say that what stood out in his mind was that I went back to arithmetic and straightened out their backgrounds first. I had assumed that *once the background was corrected then the student's natural ability would blossom.* I also assumed that *success was cumulative*, so each section we mastered incited their desire to conquer the next one.

Both students passed; hence, we accomplished our goal. We also showed that *there was great latent potential in these African American students, but we never shared this success with the administration of the University of Delaware.* Nevertheless, it shaped my thinking for carrying this successful effort to the next level.

Importance of Blackboard as a Teaching Tool

I spent 20 years working at the E.I. Dupont De Nemours & Co., Inc., where I had the opportunity to travel globally to teach various aspects of technology to customers and to Dupont overseas personnel. I wrote internal research reports and trade publications. I also gave numerous internal and external trade presentations.

Teaching technology to Dupont's global family or to customers at their mills afforded an opportunity to test ideas on good teaching methods for explaining scientific material to a cross section of people with varying backgrounds (from operators to engineers and scientists). This experience corroborated *the significance of using the blackboard as a teaching tool* when one was available.

The blackboard brought out people's weaknesses where they could be corrected immediately. It also gave the observers the opportunity to learn from the efforts of their peers in a nonthreatening atmosphere. Therefore, I found myself preparing presentations as if I were going to lecture a course in a classroom. This lecture concept often went over well with customers because I *focused my attention on these customers understanding the new technology and being able to use it immediately.*

Chapter 2

Testing Initial Teaching Premises

Opportunity to Test Some Beliefs in a High School Setting

In the early 1970s, Doretha C. Jordan, a high school English teacher in Kansas City, Missouri, asked if I would speak to a couple of English classes (D. C. Jordan, personal communication, March 5, 1999). At the time, I was writing a weekly column for the *Kansas City Globe* and other newspapers throughout the United States. I assented to her wish, looking upon this as a good opportunity to learn something about high school teaching.

    As we stood in the hall awaiting the change of classes, I was taken aback to see through the open door that the teacher was only teaching to about six students out of roughly thirty in the room. There were clusters of students doing whatever they desired at the moment. This classroom decorum ran counterpoised to my personal high school experience. I had gone to Salesianum High School in Wilmington, Delaware, an all-boys Catholic school. Salesianum has a strict disciplinary code, and students paying attention is the norm.

    One chap in this Kansas City high school class had his chair in the aisle facing perpendicular to a young lady's desk that was in the regular column alignment. This young lady displayed numerous marks on her neck as though deliberately advertising her sexual promiscuity. The teacher could have well been on another planet as far as these two lovebirds were concerned because they appeared enthralled only with each other.

I could not understand how a teacher could face this classroom disarray day in and day out and be satisfied with his or her career. My dander rose increasingly as each second ticked away whilst I stood peering into that classroom watching these children being written off. I felt compelled to do something to give these students a bit of self-respect.

I knew that if I were to have a productive discussion with these hard core students I needed to get their attention quickly. They needed to understand that I had not written them off and I expected a great deal more than what I was seeing through that open classroom.

After my introduction, I walked over to the two lovebirds. I looked directly into the young lady's eyes, so she would not only hear my words but also read my emotions. I explained to her how she really looked. I suggested that her demeanor was that of a promiscuous woman and the chap would be degrading her to his homeboys later on that day. My comments were very biting and meant to rattle her sense of self-respect. She listened to what was said and she moved her desk away from her purported lover and began to pay attention.

There was a cluster of about eight homeboys in the center of the classroom who were keeping themselves occupied by doing their own thing. I walked up to them and asked one of these chaps a question and another fellow answered it. I asked a second question of this same chap and his same homeboy answered again. Whilst looking into the eyes of the answering chap, I retorted, "He is the ventriloquist and you are the dummy!"

I had these homeboys' attention, so I made a short speech. "Education is the only thing that no one can take away from you. I can take your car, your woman, and your money."

The leader of these homeboys spoke up saying, "You can't take my woman! I'll go upside her head!"

I asked him to follow me across the room over to the windows. It was a spring day and the sun was shining. We looked down upon the grass that appeared to have been cut in the not too distant past. We could sense the new life that spring portends each year starting to blossom.

I said, "What do you think that young lady is going to be doing while you are going upside her head? She's going to bust a cap in you. Don't you know that grass turns green every year because people make excellent fertilizer?"

You could hear a pin drop in that classroom. When this young fellow returned to his seat, we started to discuss writing. My teaching style is to engage the students in a high-energy class discussion where we exchange ideas. The hour went by very quickly. Four students came up to me after class expressing a strong desire to become writers some day.

When the second class arrived, I only needed to defrock one student and everyone else fell in line because they had heard I was tough. We also had a good time discussing writing.

In pondering this high school speaking experience, I decided that clearly the teacher must command the students' attention and maintain control of his or her classroom at all times. Teachers should also keep the classroom lively enough so that they do not bore the students with the lessons they are presenting. Furthermore, *students blatantly ignoring a teacher is a form of teacher intimidation and it should be confronted immediately because left unchallenged the teacher's academic year is a write-off in his or her student's eyes.*

Testing Initial Conclusions on Teaching

Whilst I had a lot of experience teaching technology at the Dupont Company, it was generally to motivated people who either were college graduates or had received some college training. However, there was the need to know how my conclusions would impact hard-core people out of today's 'hood.

Delaware State University offered a couple of opportunities to test my conclusions and some new teaching hypotheses. I got to teach three mathematics courses at Gander Hill Prison in Wilmington, Delaware, in addition to regular college courses. At the prison, I taught regular college mathematics and a two-semester sequence of precalculus and I also was teaching these same courses to non-prison students. Thus, I had a test and control population to view my teaching methods.

DSU offered a second opportunity where I had a chance to guide the development of the mathematics portion of a joint effort with a national bank that was developing people from inner city Wilmington, Delaware, for industrial employment. The bank also used this program to upgrade the skills of their own employees. This program offered the opportunity to teach high school dropouts, high school graduates with no college training, and people who had college credits but needed a fundamental understanding of basic mathematics or first year college mathematics.

Teaching Premises

In each of the above teaching efforts, I used a set of premises to guide my mind-set in instructing these courses. They are as follows:

- Success is cumulative.
- Teacher respect must be maintained all of the time.
- Students have talent that must be cultivated.
- Everyone must go to the blackboard without the benefit of notes to show what he or she knows.
- Students going to the blackboard will find incremental success and it builds their self-confidence.
- One cannot sit down from the blackboard until he or she understands the point under discussion because this must be a positive experience.
- Students shall not debase themselves or express any feeling of doubt about their ability.
- Students cannot eat or hold sidebar conversations during class time.
- Students cannot harass each other during class.
- Understanding is very addictive.
- If students are encouraged to manage their personal problems, they will not be overwhelmed.
- If students are placed in the proper level courses, they should avoid being overwhelmed by their course load.

Premises are assumptions on which to get tests running; hence, their true value comes when they are put to the test and they prove meaningful. We shall valuate our premises by highlighting my teaching efforts at the Gander Hill Prison in Wilmington, Delaware, and a DSU/ major national bank joint program.

Gander Hill Prison Teaching Experience

When I showed up at the Gander Hill Prison in Wilmington, Delaware, I got an eerie sensation as the prison doors opened and I found myself following a guard down a hall leading to the classrooms. As we walked, I was impressed at how well this prison was kept. The good appearance made me feel good about the mind-set of those folks responsible for its operation. My apprehension over potential problems with inmate students quickly dissipated.

The classrooms were excellent. They were well lighted and they had excellent blackboards on which to work. All of the students had books and everyone was ready to get down to business.

I shall discuss the general mathematics level class experience first. This general mathematics class had twenty-plus students. There were black, Hispanic, and white students. They came eager to learn, so I did not need to spend any time attempting to motivate these students. The inmates were also accustomed to following the dictates of prison life that meant that discipline was a minor issue.

We established that everyone had to go to the blackboard and they were not permitted to sit down until the lesson was understood. Work at the blackboard constituted in essence an oral grade that was 25 percent of the final grade. There would be at least four tests of which I would discard the lowest grade in figuring out the final grade. *This dropping of one test scheme was utilized to avoid the inmates giving up due to compounding failure if they scored poorly on the first examination before I had a chance to build up their academic confidence.*

As the class got underway, it quickly became obvious that I needed to teach reading, cover arithmetic (addition, subtraction, multiplication, and division of whole numbers and fractions) and discuss decimals. Although I had apprehension over covering the material that constituted a good course, I recounted my experience with the two black students at the University of Delaware. That gave me the confidence that we would succeed.

We started by making sure everyone knew the times tables. Everyone understood that learning the times tables was necessary and there would be no compromise on this assignment. I made it clear that knowing the times tables is the basis of multiplication and division.

When we studied fractions, two key issues drove everything else. There was a need to explain what a fraction represented and the concept of a common denominator.

We first attacked fractions by giving everyone the correct terminology of numerator, denominator, and quotient. I then wrote a fraction on the blackboard and highlighted each part of it with a name. In the adjacent equation, I pointed out that 12 is the numerator, 4 is the denominator, and 3 the quotient. These proper names increased their vocabulary, and I started to attack the very serious problem of teaching these inmates to read and comprehend a mathematics book. The students then learned that the denominator tells us how many of these items it takes to make a whole item.

$$\frac{12}{4} = 3$$

Nevertheless, many students were still struggling to make a visual connection with the meaning of the denominator. I looked for a better method to get this point across and I found an excellent one chatting with the fellows standing around the street corners in Wilmington's 'hood. These 'hood fellows suggested that every example that I wanted to use be couched in terms of money because money transcends all racial and ethnic peoples and it would hold everyone's attention. Thus, the fraction concept became relevant when we talked about a quarter of a dollar meaning that it took four quarters to make a dollar, and half-dollar meaning that one must have two of them to do the same.

The idea of adding and subtracting fractions with a common denominator initially appeared to be a foreign concept, but if you couch this idea in the mind-set of money, it is an everyday occurrence. I might pose a situation where there are three pennies, two dimes, and two quarters and ask the students, "Count the coins on the table!"

"Seventy-three cents," is the usual reply.

I will counter, "There are three pennies, two dimes, and two quarters for a total of seven coins." I point out that I want them to learn to respond to exactly what is asked to be done and not what they surmised was asked. I want the students to understand how mistakes are made when we do not fully listen to what is asked.

I continue, "Tell me the value of coins on the table."

They will reply, "Seventy-three cents."

I then ask them to explain, "How did you come up with seventy-three cents?"

Testing Initial Teaching Premises

They start groping for answers. "You know you got a quarter plus a quarter and that gives you fifty cents and two dimes are twenty cents, and three pennies make seventy-three cents," is a typical response.

I challenge them to explain exactly how they came up with the seventy-three cents. My goal is for the students to learn to think about things in their most fundamental terms where true comprehension takes place.

After a couple of "you knows," I finally hear, "The quarters are twenty-five pennies each and dimes are each equal to ten pennies. You add up the pennies."

I reply, "You couldn't count how much money you have until you converted everything into pennies. Therefore, you already know the concept of a common denominator. I am not going to teach you anything new. I am only going to show you how to find the common denominator quickly."

The above exercise takes away the mystique associated with the common denominator. Students accept the common denominator in fractions as merely a different form of finding something they use in their everyday transactions. I reinforce the relationship between the concept of a common denominator in monies and fractions by posing problems such as, "You have 366 dollars to be split equally among six people, how much does each person get?"

They will say, "61 dollars!"

Again I asked the inmates to explain how they arrived at their answer and what does it mean? Should they pause too long, I guided them through understanding that it means you have six equal piles and 61 dollars in each drawn from the original 366 dollars. I go on to say that each pile represents one-sixth of the total amount because you now have six equal piles.

As we covered the arithmetic and people began to understand, they became enchanted with the power of learning. You quickly conclude that *understanding has an addictive impact on these inmates.* The more they learn the more they want to learn.

As we transition to algebraic concepts, I brainwash folks with the idea that "algebra is nothing more than high-powered arithmetic." Abstract letters used in algebra allow us to solve every problem we can imagine by merely plugging in numbers for the letters and doing the arithmetic that follows. I also made the algebraic saying become a creed that I would adopt as a guiding principle for all of the courses I would teach from then on.

In the transition from arithmetic to algebra, it seemed to work best when a student was asked to go to the blackboard to do a fraction of the form $\frac{1}{3}+\frac{1}{4}=\frac{4}{12}+\frac{3}{12}=\frac{4+3}{12}=\frac{7}{12}$. Then I challenged him to look at the fact that 12 is equal to 3 times 4. This offered the opportunity to rewrite the problem in a different form: $\frac{1}{4}+\frac{1}{3}=\frac{3}{3}*\frac{1}{4}+\frac{4}{4}*\frac{1}{3}=\frac{3}{3*4}+\frac{4}{3*4}=\frac{7}{3*4}$. In this new form, we can readily make the transition to an algebraic expression that solves all problems of this form. We generate $\frac{1}{a}+\frac{1}{b}=\frac{b}{b}*\frac{1}{a}+\frac{a}{a}*\frac{1}{b}=\frac{b}{ab}+\frac{a}{ab}=\frac{b+a}{ab}$ following the same logic that we used with the numbers under the condition that a and b not be equal to zero.

The key trick here is that we find a common denominator by multiplying the respective denominators together when we do not know how to guess this number. We also multiply the numerator and the denominator of each fraction by whatever number it will take to generate the combined denominator. We are merely multiplying each fraction by 1 because we know 1*a equals a, and $\frac{a}{a}=1$ when a is not equal to zero.

I point out here that although this work is done for addition of fractions, the same rules apply for subtraction. We take care of subtraction sign differences in doing the homework assignments.

At this point, I stress that "I am teaching mathematics and not arithmetic because arithmetic only solves one problem at a time and we want to solve all of them." Their attention to detail must also start to develop here, so I point out that the denominator cannot be equal to zero because that would result in the quotient being undefined. This attention to detail prepares them to be able to read definitions in their mathematics textbook and not overlook important mathematical details.

It was fun to watch the inmates use the derived algebraic expression to call out the sum of two fractions of the form $\frac{1}{a}+\frac{1}{b}$. However, this mathematical confidence led to the need to explain how to handle the generalized fraction of the form $\frac{a}{b}+\frac{c}{d}$ where b and d are not equal to zero.

Testing Initial Teaching Premises

A good approach was first to have a student work out on an arithmetic problem of the generalized fraction form on the blackboard to ground the concept in everyone's mind. A problem such as $\frac{3}{7}+\frac{4}{5}$ was a good place to start. Here again, we established a common denominator and then we multiplied each fraction by 1 to generate it. For example, $\frac{5}{5}*\frac{3}{7}+\frac{7}{7}*\frac{4}{5}=\frac{5*3}{5*7}+\frac{7*4}{5*7}=\frac{5*3+7*4}{5*7}$. This concept was already familiar to these inmates from the previous lesson, so they readily accepted the abstract form $\frac{a}{b}+\frac{c}{d}=\frac{d}{d}*\frac{a}{b}+\frac{b}{b}*\frac{c}{d}=\frac{ad+bc}{bd}$. I also pointed out that subtraction follows the same rule and one merely puts a minus sign in our equation where in addition the positive sign appeared.

The understanding of addition and subtraction of fractions implies knowledge of fraction multiplication. I found that by first telling the inmates to multiply the top by the top and the bottom by the bottom helped everyone to comprehend what we were doing: $\frac{3}{5}*\frac{4}{7}=\frac{3*4}{5*7}=\frac{12}{35}$. Then I used the proper terminology by saying multiply the numerator-by-numerator and denominator-by-denominator. The generalized form is now straightforward: $\frac{a}{b}*\frac{c}{d}=\frac{ac}{bd}$, where both b and d are not equal to zero.

We can now show that the division of fractions is just getting the reciprocal and following the rules of multiplication. This time we went directly to the rule that says $\frac{a}{b}\div\frac{c}{d}=\frac{a}{b}*\frac{d}{c}$ under the condition that b, c, and d are not equal to zero. Hence, a problem of this form is a straightforward calculation. In addition, the inmate students understand the definition of the term reciprocal by seeing it put to use.

The inmates readily grasped the idea that we merely divide our numbers by 100 to get their decimal equivalents, such as $\frac{37}{100}=0.37$. In a reverse manner, we multiply our decimal numbers by 100 to get their percent equivalent (0.37 * 100 = 37 percent).

After building this mathematical base, the students are able to transition to an algebraic mind-set with very little fanfare. It was now time for the students to formally start learning how to read the textbook. The arithmetic review afforded the opportunity to help the inmates learn to pronounce various words. However, now the students needed to learn how to combine words together and extract a thought from them.

I believe you should tell the students that learning is not an easy process to furnish them a realistic assessment of the task they are undertaking. My belief on this subject is summed up in the following statement: *"The brain is like other muscles, and it hurts when you first start to use it; but it gets keener with regular exercise, making the mind perform at extraordinary levels"* (Miller, 1998, p. 38).

The goal was to significantly lessen the potential that the inmate students would give up and merely count my course as another link in their chain of failures. One disquieting trait that showed up early in observing the inmates' work was that *failure had become some students' friend and constant vigilance was necessary in order to reprogram failure out of these people's lives.* An enchantment with failure allows people to discount the requirement for hard work to gain advancement in the economic mainstream. This thinking runs counterpoised to any effort to improve the lot of these inmates.

I devised two remedies to help reprogram failure out of these inmates' psyche in class. *No one could put himself down before the class* and if he did, I stopped the class and chewed him out at that moment. *All classroom discussion had to be focused on the positive.* They learned to understand that there would be enough people in the outside world attempting to put them down, so they did not need to do it to themselves. This principle was strictly enforced.

When asked to go to the blackboard, inmates could not sit down until they understood the concept under discussion. They were permitted to take only their textbook to the blackboard because each visit to "the board" became a test for I gave oral grades. If an inmate did not understand the assignment, he had to stay at the blackboard doing problems until I was satisfied that he understood the material.

Initially the thought of going to the blackboard was a very frightening thought for some inmates. They quickly learned, however, that they were on their own and I could not be fooled into believing that they had done their homework when in actuality they had not. Hence, homework assignments were usually completed.

As the semester progressed, some inmates started to not show up for class. I immediately got the guard to call them about coming to class. My goal was to make this course a successful experience for all of the students by giving the people the opportunity to stick to something when the going got tough. I lost only one student, who some of the other inmates said claimed to be afraid of me. Nevertheless, his skills were significantly behind the class and I believe he felt great stress at the blackboard.

During the lectures, we read each definition in the lessons and we discussed it to understand what it said. This approach cumulated their vocabulary and it enhanced their reading comprehension. It was quite clear that they would never learn to do word problems if I did not teach reading.

At roughly midterm (eight weeks), students started to accelerate in their learning if they had been working hard on their homework assignments. This inmate metamorphosis seemed to shift to the tenth or eleventh week for students not working up to their full potential.

The homework assignments were always the same: odd problems through number 51. This scheme offered the students a combination of easy, middle-of-the road, and difficult problems to do. *I also hypothesized that once the students did 150 problems or more they developed an acumen for doing them.*

A holistic look at how I was teaching this course suggested that my instructional model was following an exponential curve of the form CX^a where "a" is greater than zero. Therefore, I came up with the oxymoron slogan, *I am going slow to go fast!* That is, for the first eight weeks we move slowly, but we accelerate the pace each week after that for the remainder of the semester. My acceleration was governed by the amount of complaints, if any, I got from the students because they needed to be pushed to find out what they could really do.

The goal of my tests was to assess students' understanding of fundamental mathematical principles. Because everyone knew I dropped the lowest test, they did not panic if their first examination score was low. In many cases the first examination was the lowest score, then everyone would get real serious about their studies and try their best the remainder of the semester. Some people will argue that the student should take the test over to demonstrate mastery of the material he or she misunderstood on the test. I don't retest students because I fill this gap whilst they are at the blackboard.

Now you might ask, how do I get everyone to the blackboard in each class? Not everyone goes to the blackboard, but they all must be prepared if called upon to do so. If there is a room with a number of blackboards, several students will work at them simultaneously. Those people in their seats waiting their turn will often work through their problems ahead of time even though they cannot take their work to the board. Gander Hill Prison has good blackboards, so we never had a problem.

    It is interesting to see people wanting to go to the blackboard as the semester advances. The upside of that experience is it gives the students their personal confidence and they like to show off their knowledge to other people.

    I made a concerted effort to get to know students as individuals. I called people by name by the end of the semester. This effort to get to know people encouraged one young chap to want to talk for a few moments. As we chatted, he told me that I was the first adult male he ever talked with who was not teaching him how to commit a crime. He looked like a teenager who should have been at a high school football game. I asked what he was in prison for, and a distant look came to his eyes as he replied, "Murder One."

    The classroom dynamics consisted of students at the blackboard 50 to 75 percent of the time, and my lecturing or our doing reading definitions and examples the remainder of the time. This system kept the students involved in an activity throughout the class period, which left no opportunity for people's minds to stray from boredom. On many occasions when the class period ended, none of the inmate students wanted to leave class, this confirming the addictive nature of understanding.

    On the final class day, the best student in the class wanted to chat. He told me how he had never considered anything other than crime because of the rough area from whence he had come. This chap was truly elated that he had this hidden talent and that life offered a new direction when he got out.

Summary

After watching these inmate students grow academically for a semester, it was obvious that my primary role was to give people their confidence and mathematics was merely my tool to accomplish this task.

    I did not find any of my premises to be incorrect although I did not get an opportunity to test all of them in this class.

I came to understand that I was following *the model of an increasing exponential of the form X^a in pacing this general mathematics course.* This system allowed me to first review arithmetic and then start an orderly transition to an algebraic mind-set. Whilst making a transition from arithmetic to mathematics, the necessity to teach reading was highlighted. People knew how to call words, but they lacked the ability to take this string of words and put them together to arrive at a particular emotion. By going through each definition and looking at examples, we started to address the reading problem by helping inmates to understand what the definitions were saying, seeing an application of its concept, learning to pronounce new words, and building their vocabulary. We did word problems to drive home the need to read to be able to do mathematics. Students were also required to write out all the instructions for problems and the word problems whenever they were at the blackboard.

I was disenchanted with the amount of wasted human potential in this course, people who should have been doing something productive in society. Somehow this talent needs to be harnessed before these chaps are in jail. I found that if I showed genuine interest and truly believed in their native talent in hopes of developing these inmates, they responded very positively to my actions. It was also quite clear that teaching inmates requires teachers to do everything possible to keep failure from re-enchanting people for it will make it easy for inmates to give up when things get tough. A new premise to add to our list is, *academic tenacity must become the norm in the student's psyche.*

When you look at our list of premises and the teaching methods employed in teaching general college mathematics to inmates at Gander Hill Prison in Wilmington, Delaware, you find that my approach is consistent with the ideas advanced by Gary D. Borich in his book, *Effective Teaching Methods* (Borich, 1996, pp. 13–20). I would emphasize that the teacher must always remain a pillar of strength in the eyes of the inmates or I believe they will lose respect for that instructor and his or her class will degenerate into a waste of everyone's time. That implies that there can be no silliness permitted during class periods and the instructor must always come to class with his or her lecture prepared.

Chapter 3

Testing Teaching Premises in Pre-calculus Course

Added Concerns with Regular Students

Before we embark on looking at the inmate pre-calculus course of study, we need to speak about the differences between general mathematics inmates and non-inmate courses. The premises devised for inmates also hold for non-inmate general mathematics courses. However, the non-inmate students had some added concerns that had to be addressed:

1. Many people were accustomed to coming up with creative excuses to explain their lack of performance.
2. Students were not always motivated towards excellence.
3. Grades were more important to the under thirty-year-old students than comprehending the material.
4. Students were all commuters. That meant they did not necessarily interact with one another outside of class.

There was a need for a culturing process where people had to understand that excuses did not negate the need for them to get their assignments completed. I made it clear that I do not hear excuses, I only want to see performance. This mind-set quickly took hold and very few people attempted to give excuses for nonperformance. The excuses that I received were legitimate, such as true sickness or court dates. Hence, it was clear that all future *courses must start by my laying down the law on the dos and don'ts.*

The University of California offers food for thought on the importance of giving students clear direction the first day of class. They write, "Teachers who are good managers make classroom standards and rules known to students on the first day of class. These rules are explicit, concrete, clear and functional in creating a productive learning environment. There are not too many rules —only those that are essential."

The problem of igniting a pursuit of excellence was resolved by using the blackboard in conjunction with peer pressure. Once people learned they could handle the material at the blackboard and their peers observed their progress, a natural competitive spirit seemed to kindle. Everyone wanted to do his or her best without making a big deal over their goal. My role shifted to a facilitator who kept this esprit de corps alive.

There were non-inmate students who struggled at the blackboard and advanced students would pitch in and help them to understand. I encouraged this comradeship because if these good non-inmate students can teach the concepts to someone else then they had an excellent understanding of this material. I was also interested in teaching non-inmate students the importance of teamwork and letting them know that they will often learn more from one another than they will from a teacher.

With all of the non-inmate students being commuters, I gave take-home examinations with little worry over their cheating by getting together to do these tests. I found *similar test score performance for individuals regardless of whether they took in-class examinations or take-home tests.* People doing average work continued to get average scores and students excelling continued their excellent efforts without regard to test method.

Inmate Pre-calculus Classroom Modus Operandi

The expectation of students in pre-calculus was set high at the first lecture. They understood that they were being prepared to take calculus and science courses that required the pre-calculus knowledge. This meant that the pace of the course would be as aggressive as the students could handle. Nevertheless, I followed an exponential instruction model starting with a review of fractions, decimals, and percentages.

In anticipation of poor performance on the first examination, we agreed that if a student's test score fell below 75, he could do all of the problems in the sections we tested and turn them in prior to the next examination and his test grade score would be elevated to 75.

We followed the same classroom decorum as the general mathematics students. They are: no sidebar discussions, no disrespecting your classmates, no attempting to put yourself down, and only textbooks can be taken to the blackboard when you are working on problems.

Homework assignments were always odd problems numbers 1 through 51, if problem number 51 existed, or they were to do all odd problems. Students had to be prepared to handle all problems in the homework assignments at the blackboard. They were also expected to work through even problems to demonstrate their mastery of the assigned work.

Course Outline

A review of times tables and basic arithmetic was made. I also emphasized their need to understand the order of operation for mathematical expressions. I did not care which memorizing scheme people used as long as they recalled the order of operation. Two popular schemes people employ in order to memorize the order of operation are: PEMDAS and Please Excuse My Dear Aunt Sally. The meaning of the acronym PEMDAS and the order in which mathematical expressions are evaluated is:

1. **P**arenthesis
2. **E**xponential
3. **M**ultiplication
4. **D**ivision
5. **A**ddition
6. **S**ubtraction

It is very important that students understand PEMDAS because it helps to avoid getting confused when they must manipulate complex algebraic expressions. Students often forget that they must first do what is inside the parenthesis before going to the exponential, onto multiplication and division and then lastly addition and subtraction. Furthermore, one must work from left to right within a given step in PEMDAS.

The idea in stressing proper use of PEMDAS so greatly is also to help students develop good problem handling techniques early and not get caught later having to unlearn bad habits. This strategy of helping students to develop good work habits worked well with the pre-calculus inmates and it became a future creed I use for instructing all of my courses. I frankly tell students, *we want to develop good work habits by learning to do things right the first time and not have to unlearn bad habits.*

Pre-calculus Semester I Topic Focus

In addition to reviewing the fundamentals of arithmetic, the first semester of the pre-calculus course focused on the following topics as defined by Richard N. Aufmann and Richard D. Nation:

1. Fundamental Concepts
2. Functions and Graphs
3. Polynomial Functions and Rational Functions
4. Exponential and Logarithmic Functions
   (Aufmann & Nation, 1995, pp. iii–vi)

Inmate Course Dynamics—First Semester

Five students completed the first semester in the Gander Hill Prison course. Three black and two white students made up this first course.

After the first lecture, one of the black chaps asked to drop the course because he felt he could not keep up. I challenged him to not give up until he knew what he could and could not handle. He felt my supportive emotions, so he knew that I wanted him to succeed. He agreed.

This student radiated the aura of wanting to do well. Surely the black community needed strong minds, and I was looking at a significant waste of human potential in this prison. I felt good knowing that this chap was not going to succumb to the enchantment of failure and allow it to continue to consume his life. Nevertheless, I knew I must closely monitor his progress for any signs of a loss of confidence.

Working on the blackboard ignited the inmate students' confidence, and their natural competitive spirit blossomed. The student who initially worried about keeping up went beyond the homework assignment and he was doing roughly 75 percent of the problems in each section. This chap read each definition and did each example, so he was well grounded.

In the normal course of the class, we read each mathematical definition in the lesson and discussed it. Students also were required to write out all problem instructions and problems while at the blackboard. The class learned to pronounce all of the words in the lesson to build people's vocabulary. Everyone understood that one of the goals for the course was to learn to read the book. The inmates understood that when the course ended I would be history in their lives, so they needed to be able to dig out information from their textbook making it an asset. Hence, *the focus on learning to read a mathematics textbook was maintained throughout the course.*

On the first test, one chap did not pass. He opted to work out all of the problems in the tested sections. Once he completed this work, his grade was elevated to 75. His efforts offered the opportunity to examine the impact of this grade elevating technique in helping students not to get lost from a lack of understanding of lower level material as we moved onto more advanced topics. He did not fall too far behind and he started to work with the more advanced students.

An unexpected consequence of offering this higher-level mathematics course is I came to understand that other inmates called upon these pre-calculus students to explain mathematics issues to them. I was happy because I believe if you can teach someone else you must know it first yourself. It also gave these students some stature in the eyes of other inmates.

Yet as I recount the struggling student's plight, I wondered how much better his situation could have been if I had had office hours where we could have reviewed this material in-depth.

These pre-calculus students reacted similar to the general mathematics students. The pre-calculus students corroborated my belief that understanding is very addictive. When the class period ended no one wanted to leave.

There was never any problem with people displaying feelings of doubt about their abilities. I had made it clear that if they knew everything I was teaching that they did not belong in the course. Furthermore, we were all there to learn together.

The chap who initially doubted himself did require medical treatment for an ailment. His discomfort showed through in class and it seemed to kindle subtle attacks on him from his peers. His dander rose in class and I had to pull things back in line.

Everyone passed the first semester course. We covered much of the material we intended to complete. They were looking forward to the second term.

Inmate Course Dynamics — Second Semester

The classroom dynamics changed during the second term because we had a female student from the women's prison join the class. She needed to be visible to the guards at all times. This meant we needed to rehash some key points to bring the new person up to an even keel with the remainder of the class. This also necessitated our starting a new exponential teaching curve instead of merely continuing to move up the steep portion of the one from the first semester.

We had hoped to cover:

1. Trigonometric Functions
2. Trigonometric Identities and Equations
3. Applications of Trigonometry
4. Systems of Equations
5. Matrices and Determinants

I quickly learned that we needed to spend some time discussing geometry before embarking on a study of trigonometry. This study gave us the foundation to start our trigonometric discussions such as the Pythagorean Theorem and the perimeter and area of a triangle, a circle, and a parallelogram. We needed to understand the concepts of congruence and similar and what constituted a proof.

Once this background deficiency was filled, we moved aggressively into trigonometry. One student did lose his confidence and I worked through his close associate to encourage him to return. He did return and completed the course.

While marking their papers, I was disenchanted at what I saw. I came to class and said, "Marking your papers made me sick."

The inmate students were very inquisitive about my being upset and they wanted to know why. I replied, "Because it makes no sense for such great talent to be sitting here in prison." I could see in their eyes that my comments touched them deeply.

We covered the trigonometry material well but we did not have time to do systems of equations and matrices and determinants. I was not too concerned over not covering the systems of equations because this is covered in other courses. I did feel some stress at not being able to go into matrices and determinants, but these students were capable of learning these topics on their own.

Everyone passed the course. I have run across a couple of the students since their release from prison. One tells me that his nieces and nephews ask him where did he learn the mathematics he knows when he helps them with their homework. He said, "I tell them that I just know it."

The idea behind this ex-convict's using his knowledge productively became another creed I found that helps to motivate adults. I tell my students, *"By teaching you, I get your children and your children's children because you will teach your children."* Students tell me that they are helping their own children and friends, so my belief is substantiated. I think significant improvement in public education in America can come from improving the skills of the parents, so they can help to educate their own children. This desire to develop people to help themselves is part of the teaching passion that drives me to be a teacher.

Regular Student Course Dynamics—First Semester

There were two pre-calculus first semester courses to compare against the inmate course. One class ran the normal sixteen week schedule and the second was a summer school course that ran for six weeks.

*Sixteen-Week Course***:** In the sixteen-week pre-calculus course, five of six students remained at the end of the course. Three students struggled greatly with the level of the material. Two other students excelled. It was clear that the academic rigor required in this course was beyond the background of the struggling students.

One student who struggled did exercise the option of doing the problems in the tested sections twice and earned a successful grade.

The other two struggling students did not exercise the option of doing the problems in the tested area. One just gave up.

The performance of these five students suggested that I must pay a good deal more attention to the backgrounds of the students entering pre-calculus. It also suggested that students should be grouped according to their background, so the well-prepared person does not have an overwhelming advantage on the less-prepared one. This point was evident in that one of the excelling students had taken the general mathematics course first, which plugged up the holes in this person's background and that gave this student the academic preparation advantage of having seen a good deal of the material once before.

I did not know the students' major courses of study, but there was some question if a pre-calculus course designed for mathematics and science majors is at the correct level for students in all disciplines. Delaware State University has a sequence today for business and social science majors called College Algebra, Finite Mathematics, and Business Calculus. These courses address the problem posed in having a one-size-fits-all mathematics sequence.

Six-Week Summer Course I: The second pre-calculus control course sequence was taught in two six-week summer sessions. We followed the same guidelines as were used in the inmate course. The biggest problem was having a non-air conditioned classroom in the peak of the summer. I had to compete with sounds of fans running in the classroom. A couple of students got testy a couple of times from the heat and academic pressure working on their nerves. Hence, I had to be vigilant in watching for potential crises and stopping them before they turned into incidents.

There were five students: four females and one male. I gave this class a formal grade for their participation at the blackboard. This class participation grade was 25 percent of their semester grade. My goal was to understand the impact of this higher stature participation grade on attendance because I had come to believe that if the student who is struggling misses classes then I was unable to help them because there were no office hours. While I did not take roll call, students would get a zero for missing their opportunity to go to the blackboard to demonstrate their mastery of the homework assignments.

The oral grade proved to be an effective tool at getting people to attend class and be well prepared when they went to the blackboard.

Although one student struggled throughout the semester, everyone passed this course. Their enthusiasm remained through the semester.

Regular Student Course Dynamics — Second Semester

Summer Course II: There were seven students in this class, the top three from the previous summer semester and four students from other classes. It contained one male and six females. These students were all highly motivated. However, one person did poorly on an examination and this person struggled to get all of the problems in the tested sections done to raise the test grade to a 75.

We covered the same material as was done in the inmate second semester course. I employed both in-class and take-home examinations. I gave these students an oral final examination where they were given two problems and fifteen minutes at the blackboard to do them. This oral test was graded on a pass or fail grade. Everyone passed.

I also dropped the lowest test score. As one of my former mathematics professors would comment, "Everyone has a bad day" (A. Bragg, personal communication, 1964). This dropping of the lowest test score was an excellent vehicle in maintaining the students' zeal and it encouraged them to accept tenacity as the norm. Furthermore, there were no students who dropped out of this course.

Although everyone passed the course, one person barely made it. This person did not receive a transferable grade.

Summary

Student preparation for the pre-calculus course dampened the amount of material I wanted to cover. I needed to review basic mathematics and then start through algebra. This effort to fill in gaps in the students' backgrounds before moving forward maintained self-confidence or built self-confidence in students where it did not exist.

Students were taught to read a mathematics book and dig out solutions to problems that made this book an asset to them once the course was over. This meant students had to learn to comprehend definitions and see how they were employed in the examples.

Students gained a good grasp of the material covered. Because the students learned to read and figure out things from their mathematics book by themselves, I feel they can pick up the missing sections in the original course outline on their own. Furthermore, some of this material will be covered in other courses.

There was a strong need for a course sequence for non-mathematics and non-science majors because the pre-calculus sequence I was presenting was most likely beyond the scope of the needs for nontechnical people. Delaware State University resolved this problem with their present sequence of College Algebra, Finite Mathematics, and Business Calculus.

Giving an oral grade that is valued at 25 percent of the semester grade proved an excellent vehicle for insuring class attendance and student preparation. It was also understood that one could not sit down from the blackboard until he or she grasped the material under discussion.

First semester inmates students' performance was significantly better than the sixteen week non-inmate mathematics course students, but comparable to the summer school students. Inmate students were better academically and mentally prepared for the rigors of the course and they had plenty of time to study.

The second semester summer school students were highly motivated and their performance as a group was excellent. I pushed some of these students very hard to help them accept academic tenacity as the norm and they rose to the expectation I placed upon them.

The exponential instructional model that evolved while teaching the general mathematics classes also worked well in this higher-level course. It helped to develop the slogan that governs my teaching style; that is, "I go slow to go fast." By filling in students' background deficiencies, I was able to move faster and faster as the courses progressed. Once the students got to mainstream levels of progression, they enjoyed the power associated with it.

Chapter 4

Preparing Nontraditional Students for College Level Mathematics

Wilmington Mayoral Effort to Encourage Academic Excellence

There is a crisis in public school education. All sorts of efforts are being assessed to try to fix the problem. One of the most serious trends in this educational nightmare is the acceptance of low expectation by many youth. The zeal for excellence that made America strong portends to be something that the historians label a fairy tale today.

Former Mayor James Sills of the City of Wilmington, Delaware, had tried to break this gridlock of mediocre performance during his mayoral tenure. Debra Moffitt of the *News Journal* reported, "Mayor Jim Sills had proposed giving $300 gifts to students who earn a 3.0 grade point average or better. The proposal also included $100 awards for students who improve their grades and for parents who get involved" (Moffitt, 1999, p. B1).

Moffitt offers a rationale for Mayor Sills' novel actions. "Sills suggested the academic incentive after learning the average city high school student had a 'D' average last year and fewer than one in five seniors had plans for college," wrote Moffitt.

Finally, we get some very disquieting news when we ponder the long-term implications of what the city of Wilmington is facing. Moffitt reported one last disturbing concern: "City officials have been alarmed after analyzing grades, attendance rates and dropout rates among its young residents. Last year, 680 of 3,116 high school students flunked and 230 dropped out."

If we assume that the city of Wilmington is a microcosm of major U.S. cities, then the above figures suggest serious concerns about this nation remaining competitive in our new global marketplace. Can we survive with an approximately 22 percent failure rate amongst our work force of tomorrow? What do we do with the 7 percent of people dropping out of high school today because they are giving up, these people who stand a good chance of becoming America's permanent underclass?

Community College Role in Reclamation

Delaware Technical and Community College in the state of Delaware has programs to reclaim people who have potential but may have gotten lost in the school system. I will offer some observations on one effort at Delaware Technical Wilmington Campus to reclaim students who need to improve their mathematical skills. I had an opportunity to work a semester in their math lab, where students work on a self-paced basis.

I had two groups of students. The first group was learning the most basic of the mathematics offered by the college. The second group was studying algebra two levels higher.

Delaware Technical has excellent textbooks. They do an excellent job identifying the student's background deficiencies and prescribing a program to remedy them. One option they offer is this self-paced setting where people make up their background deficiencies. Students take a series of quizzes that prepare them for a test to demonstrate mastery of various topics that they needed to address.

I had only two concerns when I worked with this system. As I helped students with problems, I worried if the most basic mathematics level students needed more from their instructor than assistance in merely passing the quizzes and later the mastery tests. I was finding that I needed to help the students to believe in themselves to dispel the feeling of despair when they realized the Herculean task that lay before them.

I also believe that students at the lower level needed a good deal of assistance learning to read the mathematics textbook. If the students focus solely on passing tests, he or she may miss the extended learning, e.g., acquiring good reading skills that are important in their ability to handle other college courses. I am not merely speaking of learning to read qualitative (literature, history, and so on) material, but also quantitative material (mathematics, science, computer technology, accounting, and so on). Reading is imperative for the student's success in today's global job marketplace because good jobs can be moved from one country to another where a better trained work force exists.

Major Banking Institution Educational Uplift Program

I had the opportunity to teach in a joint program between Delaware State University and a major banking institution whose goal was to offer the city of Wilmington neighborhood people basic skills to make them employable in the economic mainstream.

The courses were taught at the banking institution's headquarters location; that meant that the facilities were excellent. There were overhead projectors, numerous computers, first-rate textbooks, and excellent blackboards, i.e., state of the art equipment.

Neighborhood community centers recommended people as candidates for this program. All races of people were amongst the student body.

Although the curriculum's design covered the total needs of the student, including resume writing and interviewing techniques, I will only discuss the mathematics, the portion that I had some influence in developing. Since the banking institution was addressing the needs of neighborhood people, this presented an opportunity to assess the soundness of the premises I had developed from observing different student populations in Delaware State University and Delaware Technical and Community College programs. My instructional premises would be tested in teaching people with backgrounds ranging from basic mathematics up to and including first year college general mathematics in an industrial setting.

The courses were based on a twelve-week cycle. We met once a week. Everyone had to go to the blackboard. General classroom protocol included no eating, sidebar conversations, and debasing oneself or one's peers. The number one issue that raised my dander was when a student debased him or herself in class. I immediately reprimanded anyone who made this mistake.

This joint effort was replete with computers loaded with nationally recognized training materials that gave the students plenty of opportunities to reinforce or supplement the classroom lectures. There were clear syllabi made up for each section, so students knew what was expected throughout the cycle.

An exponential instructional model cX^a similar to the one developed while teaching college courses at Delaware State University worked very well with these joint program students. The bank was employing computer-aided instruction that placed students into five learning levels. However, it was quickly learned that students needed to be placed in only three levels versus the five levels offered by the computer-testing program. I rarely saw a true Level 1 under the computer testing definition. The lowest background level for most people was at the computer testing Level 2. Hence, I developed my own three levels and called them Level I, Level III, and Level V. Miller Level I included the old levels 1 and 2. Miller Level III included the old level 3 and students who tested to the low side of level 4. Miller Level V included strong old level 4 and up. Perhaps better names for Miller Level I, Miller Level III, and Miller Level V would be Level A, Level B, and Level C, respectively, to reduce the potential confusion that is inherent in my nomenclature.

When each term started, I would find some students placed either too high or too low based upon the electronic tests. I placed these people based on my own judgment. However, some people did not immediately blossom, but they started to excel soon after instruction got underway. Their excellence caused a reevaluation of their potential after a few weeks of instruction at the lower level and they were upgraded to the next level. These upgrades always occurred between Levels III and V.

It was necessary sometimes to develop special courses for the upgraded students who could not attend regular advanced level courses because of time constraints. I felt it was the teacher's responsibility to seek out the hidden talents that these students possessed and push them to higher levels. These upgraded students became symbols of success to themselves and others because they made the pursuit of excellence become the norm and other students saw the payoff in prestige for excellent performance.

The bank also took advantage of this joint education effort with Delaware State University to upgrade the skills of its employees. Some people took college level mathematics and other people stayed two terms to take Level III and advance onto general college mathematics.

Every student was given a series of computer assignments that had to be completed by the end of the term. I gave each student a minimum of four assignments to do each week. I also followed up to see that they were doing these assignments because the *computer is an excellent tool to help teach reading and get people accustomed to taking tests. I chewed out anyone who was not getting his or her assignments done on time.*

I set the tone for my courses in the instructors' orientation phase of the joint effort. I made it clear that if I could not save my students some money,

Preparing Non-traditional Students for College Mathematics

then I was wasting their time coming to my class. I also hypothesized that the average person pays a 15 percent ignorance tax because they do not realize they are paying too much for goods and services. We discussed actions such as buying name brand sugar when store brand products may be just as good. We discussed the need to use the unit price in stores as the true gauge of price differences in products in different size packages. We discussed the need to put high-test gasoline into a car when its manual indicates that regular gasoline is fine.

Since fractions were taught in every mathematics course, the common denominator was discussed using actual money. I first started the lectures by placing coins on the table and then I would ask different people to count the coins. Most of the time students would give the dollar value of the money on the table. I countered with an actual count of the coins. This made people pay attention to the question being asked rather than merely respond with the first thought that popped into their minds.

I then repeated my original request. This time the students responded correctly to the question. I countered by asking them to explain exactly how they arrived at the cents value.

Initially students struggled to explain the process they used to arrive at the money value, but eventually they realized that each coin needed to be converted into pennies before it could be counted. *Once everyone realized that this conversion to pennies is a necessary condition to sum their money, the concept of a common denominator was simple to explain.* I told my students that I would not teach them anything new, that I would only show them how to do what they were already doing faster. This exercise demystifies the concept of the common denominator.

I followed the coin demonstrations by putting two hundred dollars or more in dollar bills and coins onto the table and I would again ask different people to count the money. They would count it correctly. I would ease a few of the coins off the table and ask someone else to count the money again. Again, I would be given the correct count.

I then challenged the students to say why they did not question my removing coins from the table. What usually would come through is that most people had their eyes on the dollar bills and ignored the coins. I would have these students in awe when I would point out how much money a company garners if it gets an extra twenty cents per month from each customer in a corporation with a million customers. They quickly realized that this translated to over two million dollars. They also came to realize how they could foolishly

The above exercises make learning mathematics become a personal issue for the students. It utilizes the emotions of the soul brothers (African American chaps) on the corners in the 'hood who suggested that if we used money to explain mathematics concepts, people would listen. Therefore, I am not surprised when I see an academic metamorphosis take place in valuing mathematics as a necessary lifetime skill. I also make a concerted effort to make each new item learned relevant to today's use and not some trick students put into their lifetime bag of tricks that can be used somewhere down the road.

Miller Course Level I & III

Miller Levels I and III used the same textbook. The goals of these courses were similar. The people in Level III were expected to go further than Level I people and handle much more difficult assignments.

The core areas of learning were addition, subtraction, multiplication, and division of whole numbers and fractions followed by a discussion of fractions and percentages.

The course areas touched were:

1. Operations on the Whole Number (Bittinger & Keedy, 1995) [only for low Level I and usually skipped]
2. Multiplication and Division: Fractional Notation
3. Addition and Subtraction: Fractional Notation
4. Addition and Subtraction: Decimal Notation
5. Multiplication and Division: Decimal Notation
6. Ratio and Proportion (usually introduced to only Level III)
7. Percent Notation (standard material for Level III and basic concept taught to Level I)

People were given three weeks to learn their times tables and everyone understood that this knowledge was imperative to their learning the key aspects of the above material.

We followed an instructional formula of learning to read the definitions, discuss them and see these definitions put to use in a problem. I made it clear to these students that I was teaching them mathematics. I argued that arithmetic solves only a single problem and mathematics solves all the problems of a given family type. Everyone readily accepted this mathematical belief, so it was not difficult to keep their attention on the definitions in their textbook.

We wrote out the problem we wanted to solve on the blackboard, extracted the facts we needed to solve this problem and made pictures for the problem whenever possible. This problem solution scheme gave the students a systematic approach to solving problems and it permitted them to think about what was being asked as they wrote the problem down.

Each student had his or her own set of personal problems because the neighborhood target population was people hoping to get into the work force. It became incumbent to offer encouragement to students while they worked through their personal problems. Since personal problems are unique to each individual, it is important that you avoid getting into the midst of students' personal problems and permit this activity to be handled by trained professionals.

Since I knew that an enchantment with failure could be in these students' lives, it was necessary to try to find ways to not allow frustration to overwhelm them before their backgrounds were sufficiently improved to allow their native talent to shine through. The most effective of the ideas used in the classroom was to keep a new box of tissues on the table for anyone who felt they needed to cry. But they were still required to complete the assignment. Everyone tried his or her best to stay away from needing a tissue.

However, people signaled a lack of understanding on homework assignments by saying, "*I need a tissue today*," as they entered the classroom. This became a very efficient manner of bringing out areas of lack of understanding with the student not feeling embarrassed to ask for help. I was happy to find this unintended consequence because I knew that people coming out of the 'hood often saw as a sign of weakness the need to ask for help.

The attitude problem that I saw with the Delaware State University student when I taught in 1969 reared its ugly head again with one individual. It became necessary to stand tough and not succumb to student intimidation. By the end of the term this person's mother recognized what I was doing and applied pressure at home that helped in changing this very negative attitude. What I learned from this person was that this negative attitude was a ruse to hide a lack of understanding of basic mathematics. There appears to be a technique of intimidating the teacher used by students that results in the teacher passing the students. This is perhaps described as a social promotion mind-set. This student, however, did not know I was from a public housing background, so I was tough. I expected this person to learn and I did not compromise my expectation one iota.

I had a chance to chat with other students who told how they were tough when they were in high school. They used terms such as "trifling" to describe themselves. These same people became enchanted by understanding and were very cooperative.

In this joint program, the only two assessments for which I put credence in were the tests people took at the program's start and finish. Because my instruction follows an exponential curve model, at the midterm there may be little progress to show. However, the final computer test would give a holistic look at what occurred during the term.

The final test did not always show progress even though people completed their classroom and computer assignments. However, some people jumped a couple of grade levels in the twelve-week period. I left the program before I had time to understand why some people were showing significant improvement while others were not. Nonetheless, I believe it is related to their hoping for a job at the bank once the term ended and they did not see the test as important in this process. Prior to ending my tenure in this joint program, during the instructors' orientation phase, I stressed the importance of using the final examination as a vehicle to show management the amount of progress one had made.

A key teaching strategy was to get the students to do 150 problems (50 very simple, 50 of medium difficulty, and 50 quite difficult) as quickly as possible because I had observed in other DSU courses that doing this number of problems gave students a sense for how to do problems. This strategy worked well with the joint program students.

Levels I and III students were not permitted the use of calculators until the last four weeks of the term. This forced everyone to know and use their times tables and become very proficient at addition, subtraction, multiplication, and division.

Whenever someone did not show up for class, I would call upon the bank people to go after them immediately. I knew I could not help people who did not show up to class and I did not want to devalue the worth of the class by allowing mediocre expectations to cloud their vision. Also my actions sent a symbolic message that I cared about all of my students, which allowed me to push them harder without their quitting. Absenteeism was not a major crisis but it was a constant concern.

The students went to the blackboard with only their books and they could not sit down until the concept under discussion was learned. As in other DSU courses, this technique was highly successful in helping students to understand the material and gain their self-confidence. Since going to the blackboard was a successful experience, it corroborated the notion that success is cumulative and it also helped to reprogram success as the norm in many students' lives. I even gained the nickname, "Go to the Board Sherman Miller."

It should be noted that we had some people seeking a GED in class alongside people who had completed high school. The GED people worked well. The issue was making sure you had people with comparable levels of background together. If we had people with comparable preparation grouped together, we did not hold back the more advanced people nor did less-prepared students become overwhelmed by the lectures and class assignments.

Miller Course Level V

The Level V course was directed toward getting a sufficient understanding of the general college mathematics to take a spreadsheet course and to have an appreciation of consumer mathematics. Since a banking institution sponsored the joint effort, students should learn mathematics that is important to the banking industry. This upgraded the skills of people who were already bank employees, so they could participate in new job areas not open to them heretofore.

The course areas touched were:

1. Critical Thinking Skills (Angle & Porter, 1997, pp. 13–16)
2. Sets
3. Logic
4. Number Theory & the Real Number System
5. Geometry
6. Consumer Mathematics
7. Algebra, Graphs & Functions (for very advanced students)

My instruction premises also held for this course. However, I did not find any student attitude problems in this higher-level course. It was generally felt to be an honor to be operating at the higher level.

We discussed inductive and deductive reasoning to appreciate how these two types of thoughts affect our daily lives. I posed scenarios such as the following: You have a young fellow with a car parked along the sidewalk in the 'hood, let's call him Person A. A young lady, Person B, walks by Person A's car and he walks to it and hands Person B something. A gentleman, Person C, walks by and Person A goes to his car and again hands this new person something. Finally, a second gentleman comes by and Person A does the same. Who is doing what to whom?

I then asked the students to comment on this scene. Everyone inevitably labels Person A a drug pusher. Then I stunned the students by stating that this chap was not a pusher but merely someone standing on the corner who wrote down a young lady's telephone number, gave a chap a quarter to make a telephone call, and rendered assistance to still another chap. I said the police similarly arrived at the incorrect conclusion but upon investigation found no drugs in this chap's car.

Here we discussed the dangers in improper uses of deductive and inductive reasoning. The students realized that they had erroneously profiled this innocent chap as a gangster. Nevertheless, the students comprehended the concepts of inductive and deductive reasoning.

Set theory is taught from the prospectus of how it allows the bank to manage customers in its database. We discussed how combining set theory with logic allows the bank management to target-market their database of customers by extracting groups that meet a specific criterion. The goal was to learn how to ask *yes* and *no* questions to identify specific groups of people. An example would be such as customers between the ages of 25 and 35 who make over 50,000 dollars per year and who like racing cars. To that segmented group, the bank could offer vanity credit cards with, perhaps, a picture of a racing car on its face.

Students learned that number theory and the real number system offered mathematical explanations for many things they merely memorized when they studied arithmetic. They could see how to manipulate letters and solve all of the word problems of a given style by mere substitution of numbers for the letters. This also offers an excellent method to teach abstract thinking commonplace in algebra without inciting the students' concern over the relative importance of all of the letters they are being asked to use.

In geometry, the concepts of perimeter and area were discussed. An experiment with floor tiles was done to confirm an understanding of the concept of area ($A = L \times W$). We looked at the area of a triangle ($A = \frac{1}{2} B \times H$). We discussed what it means to have two things similar, and the types of triangles. A right triangle was drawn on the board and the Pythagorean Theorem used to calculate one side. Finally, the circle was discussed where the circumference of a large circular table was determined by measuring the table's diameter and making the calculation (c = πD). The calculated circumference was compared to the actual circumferential measurement. Then the area of the table was also calculated (A = πR²).

This geometry flavor offers the students a basis to understand business situations where concepts such as area are very important. Students learned to make calculations on everyday ideas themselves. They also learned that there is experimental error when measurements are made. Students who had never been introduced to geometry got a feel for what geometry addresses.

Learning consumer mathematics was believed to be the key goal in this course, since a banking institution was sponsoring it. The rudiments of finance were covered in the consumer mathematics. Areas such as percent change, percent markups on cost, simple interest, compound interest, present value, future value, and so on, were discussed. In the case of compound and simple interest, a comparison was made of returns after three years where each step was worked out in the compound interest calculation. Finally, students were taught to use the compound interest formula $A = p\left(1+\dfrac{r}{n}\right)^{nt}$ to quickly calculate the future value of their money.

These students were prepared to handle potential bank training courses where they might discuss things such as open-end installment loans, or learn to calculate a mortgage payment by using the decreasing annuity formula.

A few students completed their assignments in 10 weeks, so I started algebra, graphs, and functions with them. Since this joint course did not offer the second semester of the general college mathematics, the Level V students had to terminate their studies just when they were academically prepared to handle algebra.

In this Level V course, we also mixed in other students who were working on their GED. The GED students performed as well as everyone else.

This course allowed me to assess the potential for people doing college work and then to advise who should go on to college. Several people who were upgraded from the Level III to the Level V class became excellent students at the higher level. One woman did roughly 75 percent of the problems in the areas covered in the textbook in addition to her computer assignments. Other people went beyond their base assignments, and asked for additional problem assignments.

People who performed well were recommended for college credit at Delaware State University and a letter grade was entered into their file.

Summary

One major drawback in this joint program is, if a student has a felony conviction on his or her record, it will be very difficult for them to work in the banking industry. This has a dampening overtone for students when they know that they will never get a job at the bank. I do not believe that this dampening feeling is lessened by the bank telling neighborhood people at the outset that there is no guarantee of employment with the institution at the program's completion. Nevertheless, I ask you to ponder, what are the chances that a young African American or Hispanic male can grow up in the 'hood without some scrape with the law? Is it not very slim, which significantly diminishes the potential of these tainted-record people finding mainstream jobs in a state where financial institution employment is rapidly growing?

We may opt to discount the importance of the felony conviction in impacting our ability to upgrade the skills of the United States until we realize that there were over two million people in jail or prison in 2003 (Nation's Prison, 2003). This high number of incarcerations has a devastating impact on minority communities as is pointed out in a 1999 report by Michael Hedges. "The jail and prison population was 41 percent white, 41 percent black and 16 percent Hispanic, with the other 2 percent Asian and American Indian. Adjusted for their numbers in the U.S. population, blacks were six times more likely to be held in jail than whites" (Hedges, 1999, p. A1).

The U.S. Justice Department reports, "At year-end 2002 . . . black inmates represented an estimated 45 percent of all inmates with sentences of more than one year, while white inmates accounted for 34 percent and Hispanic inmates, 18 percent" (Harrison, Paige, & Beck, 2003).

Since it is common knowledge that the American population is growing browner, we ought not be surprised that our high incarceration rate is now eroding our national competitiveness in the global marketplace because we are siphoning off education dollars to build prisons. Manning Marable gave us plenty to ponder on this point.

"The pattern of schools vs. prisons present in New York exists throughout the country," wrote Marable. "Thousands of black and Latino young adults in California are denied access to state universities because of the passage of Proposition 209 which destroyed affirmative action. Thousands more have been driven out due to steadily growing costs for tuition and cutbacks in student loans. Meanwhile, hundreds of millions of dollars [have] been siphoned from the state's education budget and spent building prisons" (Marable, 1999, p. 25).

Therefore, as a teacher of people coming out of the 'hood, one must keep ever present in one's mind the stigma associated with having a felony conviction and how it dampens the student's desire to advance. This concern was raised for debate during my teaching tenure in the joint DSU/bank program.

My exponential teaching model worked well with all of the students in this joint DSU/bank effort. The teaching premises worked well. It became clear that a teaching recipe where lecturing never consumed more than 50 percent of the class time was very important in keeping the student's interest high. I purposely ran a very high-energy class, never allowing the students an opportunity to stray mentally from the discussion at hand. When someone appeared to be ready to stray, I offered a diversion by having them go to the blackboard and demonstrate their understanding of what was under discussion.

Students learned how to read a mathematics textbook. They also learned to handle word problems because the computer-aided assignments were all instructions and/or word problems.

Many neighborhood people gained self-confidence and would readily tell you that they had learned to believe in themselves. They saw going to the blackboard as very significant in achieving the confidence to show off their knowledge in public.

Students also found out that they could help their own children and friends to understand mathematics. I told them "If I educate you, I get your children and children's children because you will educate them." They could see that coming to fruition while they were still in the course.

The bank setting was excellent. The bank's willingness to provide all of the tools required to do an excellent teaching job was both a joy and a good symbol of their commitment to these students excelling in their studies. I wrote weekly reports on student progress and the bank people reacted appropriately. I often made my feelings known on various problems and offered recommendations, and consequently I always looked forward to teaching in this joint program.

Each student was also given his or her own calculator to keep, so they would always be reminded of what was learned in the course and have a tool with which to make decisions when they were purchasing items or planning events.

Chapter 5

First Semester Pre-calculus Under a New Paradigm

We do not want to fully slant pre-calculus to a specific student group because it is the basic knowledge required to succeed in many disciplines. Nevertheless, we should want to find ways of getting this information over to students who have heretofore shunned mathematics and now see themselves as victims of the technology age.

A degree in business administration at many institutions today may require that one has an understanding of some business calculus and statistics. It should come as no shock to future teacher candidates that as the present national push for quality public education escalates, some understanding of calculus may differentiate many teachers of mathematics and science and govern their professional growth. But to understand calculus one needs to have a firm grasp of algebra and trigonometry. However, business calculus can be learned without a detailed knowledge of trigonometry.

The above comments suggest that first semester pre-calculus needs a great deal of emphasis placed upon the students comprehending the material and not merely being exposed to it. It is with this need to understand algebra now required of the nonscience and nonmathematics students where I place my effort in teaching first semester pre-calculus today. *The key problems I see in teaching the first session of pre-calculus are helping students learn to read a mathematics book and helping them learn to think mathematically rather than arithmetically.*

I will therefore discuss some difficult concepts for many non-science and nonmathematics students in the first semester pre-calculus to grasp and share some techniques I use to teach these principles.

I find it helpful at the first class to discuss the difference between mathematics and arithmetic. You can attempt to point out how in arithmetic you only solve one problem at a time and in mathematics you solve a family of problems, but this discussion may fall on deaf ears because many students may have no visual concept of what you are saying. I, therefore, teach this course similar to an experimental course versus a solely theoretical one.

Times Tables

I start by getting a couple of students to go to the blackboard to write out the times tables from one to twelve. I find many students may lack an understanding of the times tables and they never really memorized them during their precollege education. Therefore, the people in their chairs have an opportunity to copy from the board elements of the times tables that they do not know without exposing their lack of knowledge. I make it clear that everyone has two weeks to know the times tables.

Table 5.1 provides a basis for helping people learn to think mathematically. A first learning step is for students to understand that *multiplication is nothing more than fast addition.* Five times three equals fifteen because we have five piles each having three items. This abstract concept phrased thusly provides a mental picture for each individual multiplication effort, making the times concept easier for the students to remember. It also brings immediate utility to this concept.

You may find that earlier in their lives many students memorized a times table song yet they are unable to give you the value of a specific product such as nine times seven. The disquieting issue is that many students learned to use a calculator to do multiplication problems and never really learned the times tables. This hole in the students' learning, I believe, hampers their ability to comprehend fractions, decimals, and percentages.

I take the students through a series of observations on the times tables and offer the abstract definition that captures that point. If we examine Figure 5.1 (Five Groups of Three), we see that we have five groups of three dollar signs. This gives the students a pictorial meaning of five times three. You can use coins on a table and let people count them to achieve this same

First Semester Pre-calculus Under a New Paradigm

objective. *It is important to use a medium to which the students can relate. My experience has shown that money is an excellent vehicle for you to achieve this objective.*

Let us closely examine Table 5.1 (Partial Times Tables). The students will note that 5 x 0 = 0 and 12 x 0 = 0. If the students examine the entire times tables, they note that any number times zero gives zero. I tell them we can capture that idea by saying that a x 0 = 0, where the letter **a** represents any number. Then I follow up by asking something like *what is 500,000 x 0?*

Table 5.1 Partial Times Tables			
1 x 0 = 0	**5 x 0 = 0**	9 x 0 = 0	**12 x 0 = 0**
1 x 1 = 1	**5 x 1 = 5**	9 x 1 = 9	**12 x 1 = 12**
1 x 2 = 2	5 x 2 = 10	9 x 2 = 18	12 x 2 = 24
1 x 3 = 3	5 x 3 = 15	9 x 3 = 27	12 x 3 = 36
1 x 4 = 4	5 x 4 = 20	9 x 4 = 36	12 x 4 = 48
1 x 5 = 5	5 x 5 = 25	9 x 5 = 45	**12 x 5 = 60**
1 x 6 = 6	5 x 6 = 30	9 x 6 = 54	12 x 6 = 72
1 x 7 = 7	5 x 7 = 35	9 x 7 = 63	12 x 7 = 84
1 x 8 = 8	5 x 8 = 40	9 x 8 = 72	12 x 8 = 96
1 x 9 = 9	5 x 9 = 45	9 x 9 = 81	**12 x 9 = 108**
1 x 10 = 10	5 x 10 = 50	9 x 10 = 90	12 x 10 = 120
1 x 11 = 11	5 x 11 = 55	9 x 11 = 99	12 x 11 = 121
1 x 12 = 12	**5 x 12 = 60**	**9 x 12 = 108**	12 x 12 = 144

I have the students look at 1 x 5 = 5, 5 x 1 = 5, and 1 x 12 = 12, 12 x 1 = 12. This is captured in the mathematical rule that 1 x a = a and a x 1 = a. Here I pique the students' thinking by saying this rule does not say what the number "one" looks like.

I point out that 5 x 12 = 60, 12 x 5 = 60 and 9 x 12 = 108, 12 x 9 = 108. Here I want the students to understand that the order of operation is not important in multiplication. We capture this emotion by a x b = b x a. This presents an opportunity for the students to use their calculator to

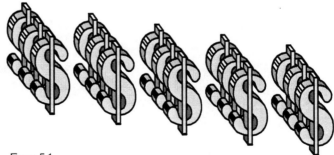

Figure 5.1

gain a bit more confidence in the fact that ab = ba. I give the students a problem like "use your calculator to see that 657 x 456 = 456 x 657." This exercise instills in the students' minds the insignificance of order in multiplication. Hence, I ask the students to ponder which of the times tables from one to twelve is the easiest to memorize? When they think about it they conclude that the twelve times tables is the easiest because you learn only one new element; that is, 12 x 12 = 144.

Real Numbers

In teaching students to read their textbook, you will need to take a few moments to discuss natural numbers as being the counting numbers, whole numbers as being the counting numbers plus zero, and real numbers (include rational and irrational numbers) are negative numbers, zero, and positive numbers. You want to stress this definition because many times they will encounter mathematical definitions that may say *n is an element of the real numbers.*

Rational and Irrational Numbers

I have students go to the blackboard to convert fractions such as $\frac{1}{4}, \frac{1}{5}$, and $\frac{1}{3}$ to their decimal equivalents $0.25, 0.20,$ and $0.\overline{33}$, respectively. These calculations offer you the opportunity to define rational numbers—numbers written in the form $\frac{p}{q}$ where q is not equal to zero and their decimal equivalents that are terminating or repeating. It is important that you show other cases such as $\frac{23}{45}, \frac{34}{200}$, and so on, because you want the students to have this background when you discuss rational polynomials that have a similar definition.

Once you have discussed rational numbers, you can state that irrational numbers are ones that do not repeat, such as the $\sqrt{3}$.

Prime and Composite Numbers

I find it instructive to tell the students that we need to examine issues from their most fundamental level if we are to truly understand what is going on. I continue that *Prime Numbers* are the most fundamental numbers in which to work because they can only be divided evenly by themselves and one. Students need to rattle off these numbers, so you are sure they have the concept.

Once students have a feel for prime numbers, the *Composite Numbers* definition readily follows: "numbers made up by multiplying prime numbers together."

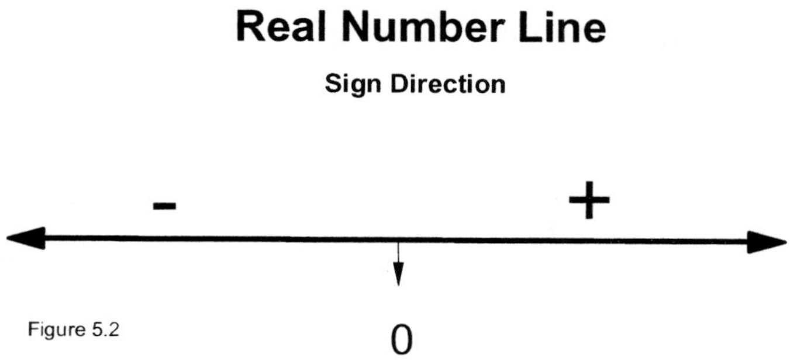

I do not spend a lot of class time having students learn to make trees to decompose composite numbers into prime numbers, but I do put a couple of examples on the blackboard to show the process. My hidden agenda here is to prepare the students for learning to factor algebraic expressions, which are processes that are comparable to decomposing composite numbers. I make this point to the students when I start to discuss factoring.

Real Number Line

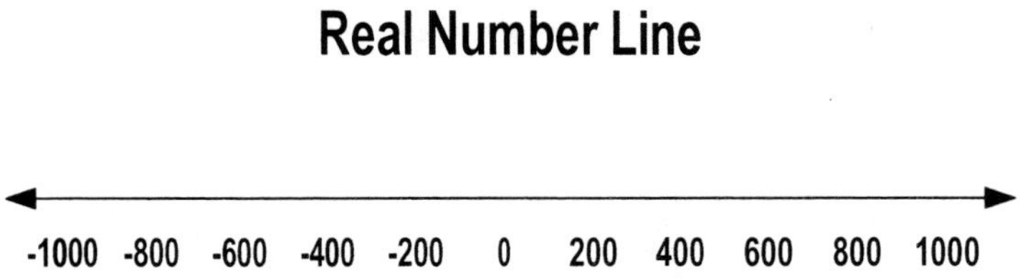

Real Number Line

Sign Direction

Figure 5.2

A discussion of personal finances offers a working picture of the Real Number Line. You might pose the situation where John wants to collect the fifty dollars from a friend who owes it to him. The friend has exactly fifty dollars in her pocket and she gives it all up. Ask the class what remains in the friend's pocket. Of course, the students will tell you, "nothing."

Real Number Line

-1000 -800 -600 -400 -200 0 200 400 600 800 1000

Figure 5.3

You cite a second situation where the friend has forty dollars when John asks for his money. Again, the friend gives John all of her money. Ask the class to explain this one. You will get something like, "the friend still owes John." You want them to say how much is left in the friend's pocket. The class will see that the borrower is still down ten dollars.

In a third situation, the friend has seventy-five dollars and gladly pays John his money. Here the class sees that the borrower has a surplus after she pays her bill.

You want the class to make pictures to capture all three situations. The key elements here are students need an equilibrium point (owe nothing) and directions that indicate when they owe money and when they have money remaining after a transaction. Figure 5.2 (Real Number Line —Sign Direction) shows that the Real Number Line captures these key items. You want to discuss these needs with the class before the students set out to draw a number line. A good scenario is to have the students devise the number line in order to resolve the above situations.

Figure 5.3 (Real Number Line) is a pictorial representation that you want to etch into the students' minds. You want to point out that the absence of a sign on the items to the right of zero means they are positive and not writing the positive sign is a world tradition. Therefore, Figure 5.3 is a quick reference point that you can draw upon when you see students' faces appearing blank as you discuss negative and positive numbers. I like to use real number lines with scales reaching into the hundreds because many students think in terms of hundreds of dollars for their day-to-day purchases.

Order of Real Numbers

Since you have established negative and positive numbers, you can now use this knowledge to show the order of numbers. If we have two real numbers c and b, then we know c is less than b if b minus c is positive. Here you want to remind the students that greater than zero means positive but not including zero. Furthermore, instead of writing out the words *less than* we will use a shorthand symbol to denote our wishes; e.g., b - c > 0 tells us that b > c (b is greater than c).

Also, if b minus c is less than zero (b - c < 0), then c is greater than b (c > b).

INEQUALITIES

I like to offer students some food for thought in the discussion of inequality because I feel this branch of mathematics greatly impacts our modern way of living. In everyday life many companies, both large and small, use databases to target-market various populations. One has to learn how to ask precise questions to make these database tools useful. Mathematical inequalities offer the key tools for operating our databases.

It is easy to give a book definition of inequalities and the students get lost listening to you. These concepts need some everyday examples to offer the students a pictorial idea of what they indicate. I believe that understanding inequalities is fundamental knowledge necessary to be able to comprehend other definitions. I will give typical examples for each inequality symbol.

1. If you are running a business and you need to find all of the people (**p**) over forty-five years old, how do you write that mathematically? The students should establish mathematically what the inequality or expression looks like ($p > 45$). You want to highlight the point that this does not include the people whose age is forty-five.

2. If you are running a business and you want to target all of the people (**p**) thirty-five years old and older, how do you write this expression? Students should be encouraged to develop the response. Here you want to point out that you included the endpoint; that means your equation is $p \geq 35$.

3. If you are running a children's store and you want to target all of the families (**f**) with members under age fourteen, you have the students recognize that this request is similar to the thinking in studying *less than* situations where the limit value is not included in the set. This tells us that $f < 14$.

4. If you are running a promotion for senior ladies (**l**) at your dress shop, how will you identify your customer base to send a flyer to these customers? Here the students are asked to define the concept of a senior lady, then write the equation. You want to see if students can quantify your request or if they need to ask questions for clarification. The students might offer you an equation such as $l \geq 65$.

In the above cases, we looked at one-sided limits. In the real world of business, we often look at two-sided limits. If a corporation is to target-market its customers, it must be able to define the criteria to extract the high potential customers from their internal database or from rental databases. Such criteria may be all customers between the ages of twenty-two and thirty-five. Here the students should see this expression as $22 < x < 35$.

I find that many students do not readily see an inequality as $22 < x < 35$; hence, it is necessary to give them a manner to read it. I use the following technique:

a. I encourage students to cover up the < 35 portion with their hands and read the single-sided inequality where they see that the goal is to find values of $x > 22$ and not including 22. The students are now looking for only values over 22.

b. Students should follow a similar use of the hands technique with the lower limit of the inequality. That is, you want to cover the lower limit $x < 22$ and read the upper limit as a singular inequality $x < 35$.

c. Have students read double inequalities aloud to ensure the concept is appreciated.

You want to stress learning these double inequalities because they appear in such places as spreadsheets, databases and computer programs.

Polynomial Definition

You can get the standard definition of a polynomial out of many algebra books. We usually talk in terms of it being as $a_n x^n + a_{n-1} x^{n-1} + \ldots a_1 x + a_0$ where $a_0, a_1, a_2, \ldots a_{n-1}$, and a_n are coefficients and $a_n \neq 0$. The degree of the polynomial is n, with n as a whole number.

It is easy to see a polynomial as a sum of monomials. This point can be highlighted in examining the individual terms in the above polynomial that is made up of various monomial terms.

Although we offer the above polynomial definition, I find that it is important here to help the students see that you only add or subtract terms with the same degree. Many students want to add the power of the polynomial when they are adding two polynomials. This suggests that you want to see that everyone clearly understands that you only add or subtract the coefficients on like degree expressions.

Given two polynomials such as $6x^3 + 5x^2 + 3x + 5$ and $3x^3 + 9x^2 + 8$, you want to show the students how they are added:

$$\begin{array}{r} 6x^3 + 5x^2 + 3x + 5 \\ 3x^3 + 9x^2 + + 8 \\ \hline 9x^3 + 14x^2 + 3x + 13 \end{array}$$

You might get each student to do one of these problems following your instruction to gain their confidence prior to going home to work on the homework. Some instructors may argue that the homework is to strengthen the students' understanding, but you want to strengthen students' confidence by offering them clarity on the material you are presenting. You want to bear in mind that in the past many business students did not need to prepare themselves to handle business calculus in college, so their goal may have been to pass only college mathematics and not a pre-calculus level course. This suggests that *you must look for ways to handle a vastly larger and diverse student population in the first semester pre-calculus course.*

What I am suggesting is that you may not need to go through the details of adding polynomials with a student population skewed toward the study of mathematics and hard science who have seen some or all of this material in high school. On the other hand, many business and social science majors may benefit significantly from you discussing these issues at a more fundamental level. Err on the side of offering greater detail unless your instincts tell you that you will appear condescending to your students.

Multiplying Polynomials

We immediately think of teaching the FOIL (First, Outer, Inner, Last) method. I find it much more instructive to teach the students products of polynomials by discussing the distributive law here. I create a couple of problems with numbers then work through some special products such as sum and difference, square of a binomial, and the cube of a binomial.

I pose a problem. Suppose you want to know what is 33 times 27 and you do not have a calculator. Of course you could always work this problem by the long method, but we are now going to use our mathematics to solve it.

1. $33 \times 27 =$

2. $(30 + 3)(30 - 3) = 30(30 - 3) + 3(30 - 3) = 30^2 - 90 + 90 - 3^2$

3. $900 - \cancel{90} + \cancel{90} - 9 = 891$

You may wish to turn the numbers around showing that ab = ba to take away any doubt lingering in the students' minds and to reinforce material taught earlier.

4. $27 \times 33 =$

5. $(30 - 3)(30 + 3) = 30(30 + 3) - 3(30 + 3) = 30^2 + 90 - 90 - 3^2$

6. $900 + \cancel{90} - \cancel{90} - 9 = 891$

Once the students see you get the same answer, you want to encourage them to work it out the long method to confirm that your answer is correct. Your goal is not merely to teach multiplication of polynomials but to also expand their ability to learn to think mathematically.

You are now ready to point out that you took the first term in the expression, multiplied it by all of the terms in the second expression, and added that to the second term in the first expression multiplied by all of the terms in the second expression. The FOIL method ought to fall out of a generalized discussion and not be seen as some special technique that the students need to memorize.

I find in working closely with students at the blackboard that many of them believe that memorizing formulas is a way to pass mathematics. I break this idea by allowing people to bring all of the information they can write on a single piece of paper to the first in-class examination. *Students quickly learn that having a formula but not understanding it means that you have nothing useful and they are persuaded to recognize the importance of comprehending the mathematics.* They acknowledge that their assist sheet meant nothing if they did not understand the material.

Students are ready to look even closer at what was done above and come up with a concept to keep in mind on how to achieve the multiplication results. Here you have the opportunity to introduce the FOIL concept. They can see what you are doing when you point out that you took the product of first terms, added to the product of outer terms, plus the product of inner terms, plus the product of last terms.

At this point I broach the issue of why it is necessary to work out each separate problem as we have done above. I ask if there is a way to do the mathematics once and merely plug in numbers to do all problems like the one above. My goal is to start students learning how to make mathematical derivations that solve all of the problems they can imagine of a given family type. Thus, I ask them to look at the product of $(x - y)(x + y)$.

1. $(x - y)(x + y) = x(x + y) - y(x + y)$

2. $x^2 + xy - yx - y^2$

3. $x^2 + xy - xy - y^2$ Here you may want to remind the students that $xy = yx$.

4. $x^2 - y^2$ The key point is that the middle terms will always disappear, so students can write the answers from their minds by merely squaring the first term and subtracting the square of the second term from it.

Students should be challenged to do a couple of problems like 38 times 42 or 93 time 87. They can go through the steps until they are comfortable that the middle terms will always disappear and that the derivation told us that fact considering that no particular numerical values are used in these expressions.

Let us follow our thinking to the next level where we want to get students able to expand $(x + y)^3$. I do not ask for $(x + y)^2$ because I use it as an exercise that I have a student come to the blackboard to go through. I broach the cubic situation with taking a number like 240 that I need to cube. We work through this problem on the blackboard using the generalized rule for multiplying polynomials. I start this discussion by

Chapter 5

reminding the students that abc = a(bc) and give an example like 2*3*4 = 2(3*4).

I then tell the students that I will write 240 as 200 + 40 to make my arithmetic easier.

1. $(200 + 40)^3 = (200 + 40)[(200 + 40)(200 + 40)]$

2. $(200 + 40)[200(200 + 40) + 40(200 + 40)]$

3. $(200 + 40)[200^2 + 200*40 + 40*200 + 40^2]$

4. $(200 + 40)[200^2 + 2*40*200 + 40^2]$

5. $200[200^2 + 2*40*200 + 40^2] + 40[200^2 + 2*40*200 + 40^2]$

6. $200^3 + 2*40*200^2 + 200*40^2 + 40*200^2 + 2*40^2*200 + 40^3$

7. $200^3 + 3*40*200^2 + 3*40^2*200 + 40^3$

8. $8,000,000 + 4,800,000 + 960000 + 64000 = 13,824,000$

Again, we want to establish a generalized form for all such problems. I point out to the students that I am merely going to replace the numbers with letters and I will follow the same procedure I did above. The key to this exercise is the following expression:

$$(x + y)^3 = (x + y)(x + y)(x + y) = (x + y)[(x + y)(x + y)].$$

The above expression and the use of the generalized multiplication of polynomials allow the students to derive an expansion for $(x + y)^3$. It is as follows:

1. $(x + y)^3 = (x + y)[(x + y)(x + y)]$

2. $(x + y)[(x + y)(x + y)] = (x + y)[x(x + y) + y(x + y)]$

3. $(x + y)[x^2 + xy + yx + y^2] = (x + y)[x^2 + 2xy + y^2]$

4. $(x + y)[x^2 + 2xy + y^2] = x(x^2 + 2xy + y^2) + y(x^2 + 2xy + y^2)$

5. $x^3 + 2x^2y + xy^2 + yx^2 + 2xy^2 + y^3 = x^3 + 3x^2y + 3xy^2 + y^3$

6. $(x + y)^3 = x^3 + 3x^2y + 3xy^2 + y^3$

If students look at step 6 and compare it to step 7 in the previous example, they will see that the terms have the same form. This allows the students to appreciate that $(x + y)^3 = x^3 + 3x^2y + 3xy^2 + y^3$ is the generalized form for solving all of these types of problems because they need only substitute numbers for the letters.

The above derivation offers an opportunity to prepare the students' mind-set to learning factoring. You want to point out that when the students see the left side of the equal sign they are also seeing its equivalent function on the right side of the equal signs. This means I can replace one function with the other.

Factoring Polynomials

I find it helpful before going into a discussion of factoring to take a few minutes to go over prime and composite numbers one more time and speak a bit about why one might be interested in the subject. The key issue is the ability to decompose things into their fundamental building blocks such as 36 = 2*2*3*3 where 36 is a composite number and 2 and 3 are prime numbers. We can say that the expression $x^3 + 3x^2y + 3xy^2 + y^3$ is a composite number and its elements are $(x + y)$, $(x + y)$, $(x + y)$, for $(x + y)(x + y)(x + y) = x^3 + 3x^2y + 3xy^2 + y^3$.

There is a need at this point to introduce the concept of finding values when something is set equal to zero known as the "Zero Property." We must first state that if **a** and **b** are real numbers then ab = 0 only if a = 0 or b = 0 or they both equal zero. Therefore, if $(x - 1)(x + 4) = 0$, then this implies that x - 1 = 0 or x + 4 = 0. Hence, x = 1 or x = - 4 because those values make the statement true. This means that 1 and -4 are solutions.

Graphs

I go further and tell the students that in the days before calculators, people needed to find ways to trace curves without making a great deal of manual calculations. We will later learn that the zero values of expressions tell

Chapter 5

us where they cross the axis of the independent variable. That says if I have an expression y = 2x, then x is the independent variable because you give a value to x and then you calculate the corresponding value for y.

Many students have plotted graphs before, so it will not be a totally foreign concept to them, but there are always a significant number of people that this becomes a first crack at really understanding the graphing technique. I, therefore, point out that a graph is nothing more than two real number lines laid perpendicular to each other where you plot the value of the independent variable on the horizontal axis and the dependent value on the vertical axis. Since a picture is worth "1000" words, I draw it out on the blackboard and plot a couple of points for the expression y = 2x. I use both negative and positive numbers and zero to show that y is zero when x is zero which tells us that the graph passes through the x axis at x = 0.

Although I will repeat this discussion later, I state without a detailed development that I can draw the line once I know a couple of points, in this case (0,0) and (1, 2) where the first term in the set is the independent variable and second term is the calculated dependent value. The idea here is that we can construct graphs by knowing only a few key points such as where it crosses the independent variable axis.

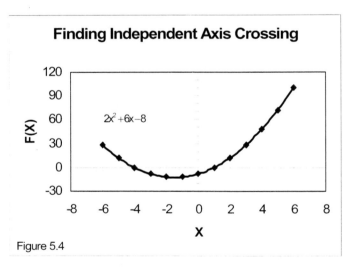

Figure 5.4

Factoring will help us find places where expressions have dependent values equal to zero which tell us that the curve crossed the independent axis at these points. I pose that if we have the equation $y = 2x^2 + 6x - 8$ we can find where it crosses the x axis by factoring it and setting each factor equal zero.

Thus, we have $y = 2x^2 + 6x - 8 = 2(x^2 + 3x - 4) = 2(x + 4)(x - 1) = 0$. That tells us that x = - 4 or x = 1 are solutions to this expressions and these points are where the graph crosses the x axis. Students need to see this plot to have confidence that indeed what you say is true. Figure 5.4 (Finding Independent Axis Crossing) shows that the curve crosses the x axis at two points and they are - 4 and 1 which corroborate your contention and it offers utility for learning how to factor.

Figure 5.4 allows you to explain Cartesian coordinates again visually because the students can now see the points (- 4, 0) and (1, 0) on the curve. This also gives the students a look at a nonlinear curve.

Going through the above graphing information here also provides an opportunity for students to allow percolation time to enhance their understanding because the material will tumble in their minds and it will not be new when you return to this discussion.

Domain

The importance of understanding the domain concept in an algebraic expression is pivotal to understanding higher mathematics. I explain the domain of a function by telling the students that "*It is all of those values where the expression makes sense.*" In more mathematical language, the set of real numbers where the function is defined.

It is very important to give several examples where the students need to figure out the domain such as:

$y = \frac{1}{x}$ where $x = 0$ is the only real number that will not work, so the domain is all $x \neq 0$.

$y = \frac{1}{x^2 - 9}$ where $x = \pm 3$ are the only real numbers that will not work, so the domain is all $x \neq \pm 3$.

$y = \sqrt{x - 10}$ where the radicand must be greater than or equal to zero to have real solutions. This implies that $x - 10 \geq 0$ and tells us that $x \geq 10$.

One of my hidden agendas here is to strongly encourage the students' budding mathematical reasoning skill. I ease the students away from the practice of memorizing without comprehension to where they are able to think their way through situations. You may see some students struggling to feel comfortable with the domain concept here, but it is very important that you give this concept the added time necessary to become a part of the students' psyche. Figuring out domains are excellent exercises for learning a multiplicity of mathematical principles, so that bit of extra time to ensure that students master this concept allows you to pick up the pace in the pre-calculus course.

Reducing Rational Expressions

It is always helpful to approach subjects in pre-calculus from the point of view of demystifying the abstractness of the material. I like to offer the students an arithmetic problem that underpins the algebraic concept I am about to present. This allows me to argue the case that "*I am not really teaching you anything new; rather, I am only showing you how to solve any problem of a similar nature.*"

Consider rational numbers of the form $\frac{p}{q}$, where the value of $q \neq 0$, that are reduced to their lowest terms. If we look at the rational numbers $\frac{1}{2}, \frac{3}{6}, \frac{4}{8}, \frac{5}{10}$ and $\frac{45}{90}$ we know they all equal the same value $\frac{1}{2}$. How we come to one-half is the issue the students need to quickly visualize. If we rewrite these rational numbers as $\frac{1}{2*1}, \frac{3}{2*3}, \frac{4}{2*4}, \frac{5}{2*5}$ and $\frac{45}{2*45}$ and use the concept that $\frac{a}{a} = 1$ (when $a \neq 0$), we see they all decompose into $\frac{1}{2} * 1 = \frac{1}{2}$. Now the students understand the reason for canceling out like things from the numerator and denominator.

I remind the students that algebra is nothing more than high-powered arithmetic, so it makes perfectly good sense that we should do similar cancellation operations with algebraic expressions. I continue that when $\frac{a}{a} = 1$ there is nothing in this expression that says what is the composure of "a." Thus this allows us to describe very complex functions to be "a" if that is our desire.

You want to start with a problem the students can recognize easily but reinforces their basic understanding of a heretofore derived expression. I like to start with $\frac{x^2 + 10x + 25}{x^2 - 25}$. It is always good to have a student go to the blackboard and you guide him or her through this mathematical calculation.

At this point, I start encouraging the students to ask if they fully understand the problem and if so do they have the resources to solve it. Students make many mistakes because they do not fully comprehend what they are being asked to do. An underlying objective here is to subtly force students to learn to think through an issue as a matter of a routine course of action and not merely work under some erroneous personal belief that does not answer the query. Students in this first course should understand that they will be paid in the global job market based on their ability to find solutions to problems that their management deems important. To do so, they must first clearly understand what is being asked. As teachers, we must look for every opportunity to teach students to think.

In returning to the immediate problem, we need to make sure that the students understand that we want them to reduce this problem to its lowest terms just as we do for any fraction made up solely of numbers. This implies that we must decompose both the numerator and the denominator and look for common terms. I suggest to the student at the blackboard that this is the time to use his or her factoring skills.

The student should offer a numerator decomposition such as what follows:

$$x^2 + 10x + 25 = (x + 5)(x + 5)$$

If he or she is struggling to come up with these factors, you should offer the student a couple of similar problems before proceeding because there would be many other students also struggling with this concept. Pay close attention to how many of these side problems are necessary for the student to grasp the factoring concept. This is your barometer for determining how successful you were in getting the factoring concept over to your class. It may be necessary to take a couple of moments and go over factoring one more time to ensure that the students do not fail to comprehend the new ideas you are presenting.

Hence, the student should offer the following for the denominator:

$$x^2 - 25 = (x + 5)(x - 5)$$

Now we want to reconstruct the original problem including these decompositions and it should become readily apparent to the students that they can cancel (x + 5) from both the numerator and the denominator as they would do in reducing any fraction.

$$\frac{x^2 + 10x + 25}{x^2 - 25} = \frac{(x + 5)(x + 5)}{(x - 5)(x + 5)} = \frac{(x + 5)}{(x - 5)}$$

You want to call several other students to the blackboard and have them work through similar problems. Once you feel they understand the concept, then you want another series of students to come to the blackboard to work on problems with a higher degree of difficulty that you devise in class. *I feel it important that you devise these problems in the classroom because symbolically you show that your reputation is as much open to public scrutiny as what you expect of the students.* This action clearly signals that you are not asking the students to do more than you would do yourself.

You would have to be concerned here with mathematics and hard science majors, but you want to bear in mind once again that you are offering higher level material to a mass audience which necessitates that you take extra effort to see that the students comprehend the lessons.

Adding Rational Expressions

This becomes a straightforward operation if one goes through a fractional addition such as $\frac{2}{7} + \frac{4}{9}$ first. The solution is listed below:

- You want to remind the students that we need to find a common denominator for $\frac{2}{7} + \frac{4}{9}$ before we can add these two fractions.

- I point out that although students heard a great deal of discussion about finding a least common denominator when they were in high school, it is only necessary to have a common denominator to solve the problem.

First Semester Pre-calculus Under a New Paradigm

We can reduce the answer to its lowest terms.

It is good to have a student do this fraction at the blackboard. You might find that many people will set it up as:

$$\begin{array}{l} \dfrac{4}{9} \\ +\dfrac{2}{7} \end{array} \qquad \dfrac{4}{9} = \dfrac{4*7}{9*7} = \dfrac{28}{63} \\ \dfrac{2}{7} = \dfrac{2*9}{7*9} = \dfrac{18}{63}$$

But when you go back and do the same problem using multiplication by 1 you may need to do some explaining because they have only memorized the above process without understanding what they were doing. I insist that they also learn to solve the above problem by setting it up as:

$$\dfrac{4}{9} + \dfrac{2}{7} = \dfrac{7}{7} * \dfrac{4}{9} + \dfrac{9}{9} * \dfrac{2}{7} = \dfrac{28}{63} + \dfrac{18}{63} = \dfrac{46}{63}$$

My goal is for the students to recognize that what they are doing is merely multiplying by 1 and the common denominator is the product of the individual denominators. *The trick in multiplying by 1 is to ask oneself what number do I need to multiply the individual fraction denominator to generate my common denominator value?* Once one recognizes that number, they make a fraction where it becomes both the denominator and numerator. The above expression utilizes this technique to generate 63 as the common denominator. In one case we need to multiply by 7; in the other we multiply by 9.

You want to keep this approach on the blackboard as you turn to an algebraic problem. Assume you ask the students to add the following algebraic fraction: $\dfrac{4}{x-7} + \dfrac{5x}{x+2} + \dfrac{9}{x-1}$. If students follow the technique established above, they should be able to tell you that the common denominator is $(x-7)(x+2)(x-1)$. You want to draw out of the students what *1* looks like in each case to get everything over the common denominator. Thus, we simplify this problem as follows:

$$\dfrac{4}{x-7} + \dfrac{5x}{x+2} + \dfrac{9}{x-1}$$

$$\frac{4(x+2)(x-1)}{(x+2)(x-1)(x-7)} + \frac{5x(x-7)(x-1)}{(x+2)(x-1)(x-7)} + \frac{9(x-7)(x+2)}{(x+2)(x-1)(x-7)}$$

$$\frac{4[x^2+x-2]+5x[x^2-8x+7]+9[x^2-5x-14]}{(x+2)(x-1)(x-7)}$$

$$\frac{4x^2+4x-8+5x^3-40x^2+35x+9x^2-45x-126}{(x+2)(x-1)(x-7)}$$

$$\frac{5x^3-27x^2-6x-134}{(x+2)(x-1)(x-7)}$$

We have gone through all of the algebra gymnastics but this expression has little value to the students unless you offer them a reason for going through these exercises other than for mental development. Mental development is great for mathematics majors but business people are much more pragmatic and require more justification.

I tell students that they will learn to use spreadsheets to be effective in today's global business environment and this expression we just worked out tells us how to add all of the number expressions of the form $\frac{4}{x-7} + \frac{5x}{x+2} + \frac{9}{x-1}$ that we can imagine where we can input various values of x. However, we are restricted to the values of x in the domain of our function that tells us that $x \neq -2, 1,$ and 7.

But the proof of the pudding is in the eating. You need to substitute a couple of numerical values in for x and solve the problems with both the derived expression and by the traditional method. Here I will look at 3 and 15 (see Table 5.2 - Standard Vs. Solution Method).

Case I where x is equal to 3.

$$\frac{3}{3-7} + \frac{5*3}{3+2} + \frac{9}{3-1} = 6\frac{1}{2}$$

$$\frac{5(3)^3 - 27(3)^2 - 6(3) - 134}{(3+2)(3-1)(3-7)} = \frac{-260}{-40} = 6\frac{1}{2}$$

Case II where x is equal to 15.

$$\frac{4}{15-7} + \frac{5*15}{15+2} + \frac{9}{15-1} = 5\frac{66}{119}$$

$$\frac{5(15)^3 - 27(15)^2 - 6(15) - 134}{17*14*8} = \frac{10576}{1904} = 5\frac{66}{119}$$

The above exercise will convince the students that you have achieved your goal. But you ought not be surprised when they complain over the difficulty in using the complete solution. I counter that with this is the equation you would write in a spreadsheet to generate a host of answers where x is given many values. Table 5.2 offers both methods used to calculate the results versus values for x and you can clearly see the undefined points. This table helps the students to understand that both techniques yield the same value and except where x equals -2, 1, and 7, all other real numbers will work in these expressions.

This is an experimental approach to teaching the definition of the domain of x. It is good to give students a similar problem and allow them to use their calculators to generate a table showing the undefined points in their expression, if there are any.

Table 5.2
Standard Vs Solution Method

x values	Standard	Solution
-2	#DIV/0!	#DIV/0!
-1	-10.000	-10.000
0	-9.571	-9.571
1	#DIV/0!	#DIV/0!
2	10.700	10.700
3	6.500	6.500
4	5.000	5.000
5	3.821	3.821
6	1.550	1.550
7	#DIV/0!	#DIV/0!
8	9.286	9.286
9	7.216	7.216
10	6.500	6.500
11	6.131	6.131
12	5.904	5.904
13	5.750	5.750
14	5.639	5.639
15	5.555	5.555

Table 5.2 was initially made in Lotus 123 and then converted to Excel. The Lotus 123 generating cell formulas for the standard and solution form results are:

Standard form: 4/(C440-7)+5*C440/(C440+2)+9/(C440-1)
Solution form: (5*C440^3-27*C440^2-6*C440-134)/((C440+2)*(C440-1)*(C440-7))

If you have not used a spreadsheet, you only need to replace the cell addresses in these equations to make things work. You put your x value at the cell address you wish to calculate and substitute that address for C440 when you type your equation into the spreadsheet. Although the present effort was done in Lotus 123, comparable equations can be developed for Microsoft Excel and Corel Quattro Pro.

Cartesian Plane

It is good to introduce the discussion of the Cartesian coordinates with an experiment where you get the concept of what is going on into the students' psyche. If you are in a rectangular-shaped classroom, you might try having a student stand at a location that is clearly visible from the perpendicular formed when two walls meet. Show the class that to reach the student you need move along one wall to where the person is on a perpendicular line from this wall to you, from that spot you move parallel to the second wall out to where the person is standing. Repeat this exercise a couple of times until people feel comfortable knowing that two moves will locate every point on the surface of the floor.

You now want to call one wall the x direction and the perpendicular wall the y direction. If you bring a tape measure to class, you can have people measure how far along the x and y walls they must go to reach various points.

You may then show that measurements along the walls are tantamount to creating two number lines that are perpendicular to each other where the horizontal line represents the independent axis and the vertical line represents the dependent axis. This means that we move along the independent axis first and it dictates where we locate on the dependent axis. Once we move down the x wall, there is no question where we must go on the y wall to reach our objective.

We now call the tape measure a scale for our walls and submit that the standard form of locating a point in this new Cartesian system is to give the value of the independent variable first and then the dependent one second. Therefore, we locate the point (3, 6) on the floor by going along the horizontal first three units and up the vertical 6 units.

We must also bear in mind that sometimes we need to move in the negative direction on our number line. Suppose our person is located at (3, -6). This tells the students that they must move in the positive direction on the horizontal line but move in the negative direction on the vertical line.

Figure 5.5 (Coordinate System) is a pictorial representation of a coordinate system. Here you can show that this coordinate system is

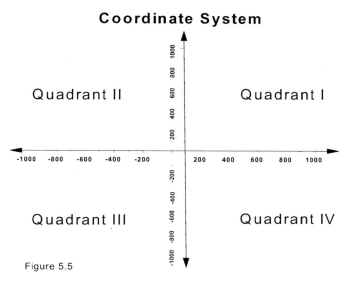

Figure 5.5

merely an overlaying of two number lines that are laid down perpendicular to each other. Figure 5.5 was made just as described above.

Pythagorean Theorem

The Pythagorean Theorem offers us another opportunity for the students to make a calculation and confirm it by a measurement. You want to get a student to go to the blackboard and draw out a right triangle. Then label its legs a and b and the hypotenuse c (see Figure 5.6 Right Triangle). Since you want to encourage the students' learning to appreciate how to think mathematically, you might state that the Pythagorean Theorem says that the sums of the squares of the legs in a right triangle are equal to the square of its hypotenuse.

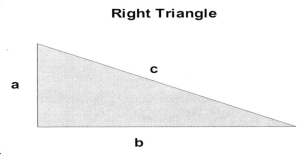

Figure 5.6

92   **Chapter 5**

The above statement of the Pythagorean Theorem then leads one to write the equation: $a^2 + b^2 = c^2$ or $c = \pm\sqrt{a^2 + b^2}$ but as the c length we take $\sqrt{a^2 + b^2}$. A second student should come to the blackboard and measure the legs labeled a and b of the right triangle and calculate the length of side c, but not share his or her results with the remainder of the class.

A third student should go to the blackboard and only measure c and write his or her results on the blackboard. Then ask the second student to place the calculated results next to the experimental results. When the students see how close these values are, assuming the measurement is made correctly, they will come to appreciate the practical utility of the mathematics you are teaching. We also want to use this discussion as the underpinning to our introduction of the distance formula.

Distance Formula

The students are learning to think mathematically and it now is time to use that ability to develop the distance formula from fundamental principles. You want to encourage the students' reasoning ability and discourage rote memorization, so you need an experiment to show students one of the key concepts in building the distance formula. That is, finding the distance between two points on a real number line. The idea you want to bring out is that subtraction finds the physical difference between two items relative to a fixed point. When we show $6 - 5 = 1$, that says if we start out at a fixed point zero and come out 6 units and then come out again from zero to 5 units. Then when we cut the five units section from the six units line, we see the line difference length of one unit is remaining.

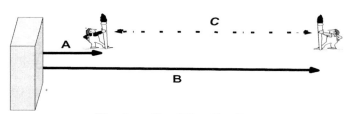

Figure 5.7

I usually get two students to stand some distance away from a wall with a parallel separation between the people. Figure 5.7 (The length of C = B - A) is a pictorial representation of the experiment that I have members of my class perform. Here students can readily see that from the fixed point of the wall they can cut A from B and end up with the distance C.

Once students are comfortable with the idea of subtracting to find the length between two positions, you are ready to work through the logic of the derivation of the distance formula.

We will call back our coordinate system and place two points in space and label them (x_1, y_1) and (x_2, y_2). The key idea here is that the line formed by these two points have cast lengths that show up on both the x and y axes. These difference lengths are the legs of a right triangle. Therefore, the students are able to use Pythagorean Theorem to calculate the length of the hypotenuse from knowing the legs.

But we need to convert these leg-lengths into values that fit into the Pythagorean Theorem. We can do that by labeling the length differences

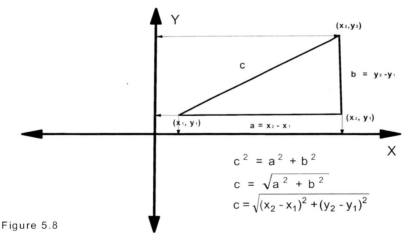

Figure 5.8

with letters. This is a process you want to walk the students through a couple of times until they are comfortable with it. The point is that we picked the two points (x_1, y_1) and (x_2, y_2) and these points create cast-differences (projections) on both the x and y axes. When we slide these cast-differences along parallel lines they form the legs of a right triangle (see Figure 5.8 Distance Formula Derived).

We can readily show the students how they can calculate the hypotenuse of the right triangle by using the Pythagorean Theorem. We know that

$c^2 = a^2 + b^2$. The students should understand at this point that they need only substitute values x and y for a and b. From this new equation the students can derive the length of the hypotenuse by taking the square root of both sides; to wit, $c = \sqrt{(x_2-x_1)^2 + (y_2-y_1)^2}$. This last expression is the derivation of the distance formula. Students should be encouraged to use it to calculate the distance between two points and also to measure the length between these points to see the utility of their derivations.

Midpoint Formula

Students can use simple reasoning to come to the midpoint on the line joining the two points (x_1, y_1) and (x_2, y_2). If one wants to know the midpoint between two numbers, one would add the two numbers together then divide their sum by two. Students can use this same mind-set to arrive at:

Circle

$$\text{midpoint} = \left(\frac{x_2+x_1}{2}, \frac{y_2+y_1}{2}\right)$$

"A circle is the set of points that are a fixed distance from a specified point. The distance is the radius of the circle. The fixed point is the center of the circle" (Aufmann & Nation, 1995, p. 96).

If you get the students to think through this definition, the equation for a circle becomes obvious. Parse the circle definition and encourage the students to write expressions for each step.

- "...Set of points that are a fixed distance from a specified point." This sounds just like what we were doing when we placed points (x_1, y_1) and (x_2, y_2) in our coordinate system and attempted to find the distance between them.

- "The distance is the radius of the circle." The students already understand this distance concept, so it is no surprise when you say that the distance formula for these two points is $c = \sqrt{(x_2 - x_1)^2 + (y_2 - y_1)^2}$. However, instead of using "c" for our hypotenuse, we will use "r" to represent it.

First Semester Pre-calculus Under a New Paradigm

- "The fixed point is the center of the circle." Since the point (x_1, y_1) is our starting point, it becomes the center of the circle. Instead of using symbols (x_1, y_1) we will use (h, k) to represent our starting point. You also point out that you are replacing symbols (x_2, y_2) by (x, y) because you are merely describing some point on the x, y plane. This means we can now write the standard form of the circle as $r^2 = (x-h)^2 + (y-k)^2$.

Here the students should be able to solve for $r = \pm\sqrt{(x-h)^2 + (y-k)^2}$.

Ask the students, "What does the above equation look like when the center of the circle is at the origin?" They ought to be able to arrive at it with little difficulty. If they have difficulty, you want to remind the students that the origin means point (0, 0).

At this point in the course, students should be starting to feel some degree of comfort with your parsing definitions and their being able to use them to derive equations versus exploiting rote memorization. Since *one of the key goals of the course is to teach how to read a mathematics book, you have achieved the underpinning of that objective at this point.*

GRAPHIC REPRESENTATION OF DATA

The old proverb, "A picture is worth a thousand words," should become a creed of students seeking to become professionals. They are expected to present their findings in formal presentations that are the accepted way of presenting material in the world marketplace. This means that students need to understand how to manage data.

Since we are discussing graphing material in pre-calculus, this is an opportunity to offer the students some aids in examining data. It is important that you not only show students how to manage the data, but you also help them to understand the story it is telling. *Industries and civic groups expect professionals to be able to interpret what the numbers say, so if you are merely teaching how to rack up the data, you have shortchanged your students.*

I like to send a student to the blackboard and ask him or her to write down two rows of numbers with ten numbers in each row. The student might give something similar to Table 5.3 (Student Generated Numbers).

Table 5.3
Student Generated Numbers

24	89	76	9	87	25	45	77	13	44
55	87	66	56	65	99	23	46	23	34

One method of racking up this data is to use "Stem and Leaf Plot." The key issue is to make sure that students understand the digits placements in a number. Consider the number 3,246. It is helpful to take a brief moment to walk the students through what each digit represents. You are better off doing this subtly where you might say, "Here we see 6 represents the ones place where you can have numbers from zero to nine. The number 4 represents the tens place. The number 2 represents the hundreds place. The number 3 represents the thousands place.

If we are writing a check for this amount, we write three thousand two hundred forty-six dollars. That is the same as saying 3,000 plus 200, plus 40, plus 6 dollars.

With the digits places in mind, you are now ready to help the students understand how to develop the stem and leaf method. The stem will host the primary unit such as 9 in 98 and the leaf will capture the various secondary value (8). A second input would be 2 for the stem and 4 for the leaf. A third input would be 7 for the stem and 6 for the leaf.

Table-5.4 (Stem and Leaf Method) tells us that 40 percent of the data is clustered between the stems of 2 and 4. Encourage the students to take a closer look at the stem and leaf technique and they will see that it is a de facto histogram. I find that I need to enhance the students' vocabulary by pointing out that histogram is the same as a bar on a bar chart.

Table 5.4
Stem and Leaf Method

Stem	Leaf
0	9
1	3
2	4533
3	4
4	546
5	56
6	65
7	67
8	977
9	9

You need to go through each element in Table 5.4 before you proceed. You might want to put a few students to the blackboard to carry out this exercise.

We move on to the bar chart. Figure 5.9 (Applegate Produce Sales Bar Chart) shows the sale (in millions) of Applegate produce over ten years. The students can see that the business deteriorated over the decade. The slight increase for 2004 may be an illusion considering the past performance of this company.

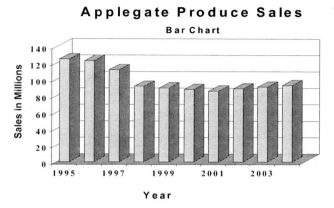

Figure 5.9

It is good for the student to see that the same data can also be reported in a line chart. Figure 5.10 (Applegate Produce Sales Line Graph) is the same Applegate produce sales reported in a line graph format where the specific data points are given. Although we can use finer gradations on the sales volume axis, our goal is to show a trend and not go into all of the minute details. Business executives are not interested in all of the glorious details that employees' data will show, so the professional's job is to supply the necessary data for his or her management to make good business decisions and not cloud the issues showing off his or her neat graphing skills.

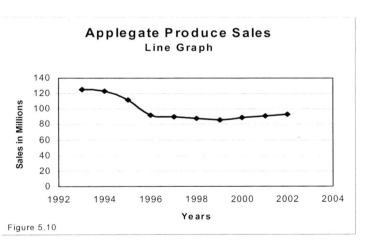

Figure 5.10

We have now completed the preparatory work needed for the students to feel comfortable learning pre-calculus. They have the tools necessary to move quickly for the remainder of the course. This means I start up the steep portion of my exponential teaching curve for the remainder of the course. It is very important here to determine how quickly the students can handle new concepts so that you can gauge how steep the learning curve is for a particular class you are teaching.

You want to bear in mind that the steepness of the class learning curve is a judgment call that you have to make as you start to move aggressively into presenting new course materials. This suggests that you find what learning rate your class can handle and teach just a bit under that to allow for potential mistakes in your own reading of the situation. It is good to have a safety factor because you do not want to reintroduce unnecessary class frustration that may encourage your students to reincarnate any dormant feelings that failure is okay or no one expected them to learn pre-calculus.

You might note what the literature suggests can happen to your class if you allow frustration to raise its ugly head. Rosalind Reed offers us sobering comments on why we want to keep frustration from overwhelming our students. Reed's descriptions includes phrases such as, "Overt hostility from a student; verbally aggressive." She states further, "[Students] usually become verbally abusive in frustrating situations which they see as being beyond their control; anger and frustration becomes displaced onto others; fear of rejection and feelings of righteous indignation are frequently associated with this pattern" (Reed, 1997).

Clearly Reed's comments suggest that you do not want to introduce problems with frustration in your class. Thus, you want to have some cushion when you start up the steep portion of the exponential learning curve model you employ for the course.

Functions with Their Graphs

We are interested in creating working definitions for people who are not going to major in mathematics and as a consequence have no need to get lost in all of the subtleties of the subject. The mathematicians will define function in two manners:

Rule Form of the Definition of a Function

"A **function** is a rule that produces a correspondence between two sets of elements such that to each element in the first set there corresponds *one and only one* element in the second set.

First Semester Pre-calculus Under a New Paradigm

"The first set is called the **domain**, and the set of all corresponding elements in the second set is called the **range**" (Barnett & Ziegler, 1993, p. 196 – 197).

Set Form of the Definition of a Function

"A **function** (input/output process a transformation) is a set of ordered pairs with the property that no two ordered pairs have the same first component and different second components.

"The set of all first components in a function is called the **domain** of the function, and the set of all second components is called the **range**."

The above definitions are excellent for mathematics and science majors, but business students seeking a working knowledge of the concept need a simpler expression of this definition. Nation and Aufmann give something a bit more palatable to business and social science students. They write, "A function is a relation in which no two ordered pairs that have the same first coordinate have different second coordinates" (Aufmann & Nation 1995, p. 96). This definition does not worry about domain and range.

Finally, let us look at one more definition of a function offered by Roland E. Larson, Robert P. Hostetler and Bruce H. Edwards. They write, "A function f from a set A to a set B is a rule of correspondence that assigns to each element x in the set A exactly one element y in the set B.

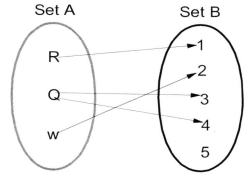

Figure 5.11

R and W represent functions

Q does not represent a function because the element Q in A has two different values in B

The set A is the domain (or set of inputs) of the function f, and the set B contains the range [or set of outputs]" (Larson, Hostetler, & Edwards, 1997, p. 103). Figure 5.11 offers a pictorial representation of this definition.

You will want to walk the students through one of these definitions where they have a pictorial representative beyond the standard graphic represented in Figure 5.11. I will use the last definition and go through it. My approach is to use a simple example, such as y = 2x + 4. The idea here is that we have a rule that says whatever real number you want to make x you must multiply this number by 2 and add 4 to it to generate the value of y. The rule simply tells you what must be done to get y.

It is good if you also give students the definition that says *y is equal to the f(x)*. At this point, you want to get the students seeing f(x) wherever they see y which should make the concept of the function a bit more palatable. This introduction allows the students to write f(x) = 2x + 4 as the true example to work out. You can also take comfort in the fact that you are laying the ground work for the students to be familiar with *y = f(x)* when they study both finite mathematics and business calculus.

We have not parsed the definition of domain in our discussion. In your pursuit to find a simple way to help the students understand this concept, you might broach this subject with a simplistic definition such as the domain being the set of independent variable inputs that make sense (where the function is defined). If you have a function $f(x) = \frac{2}{x}$, then it is clear that zero cannot be a value in the domain of x.

Encourage the students to write out the domain of $f(x) = \frac{2}{x}$ in a couple of different manners. You might expect to see responses such *as all real numbers except* x = 0, $(-\infty, 0) \cup (0, \infty)$, or simply x ≠ 0.

You must highlight to the students the need to examine each function to assess its domain. This effort encourages students to become accustomed to thinking, and it makes this domain discernment the expectation.

Graphing

You might want to remind the students of the experiment that they did in locating a person standing on the floor of the classroom using the two walls as reference locations when you were defining Cartesian coordinates. With this location picture in mind, you introduce graphing as a series of determined locations that follow a prescription given by finding solutions to an equation and then you connect the points together. It is worth reminding the students that a solution to an equation is a value that makes the statement true.

Although some instructors emphasize using graphing calculators or graphing routines, it is important at this critical juncture to not succumb to this temptation because the students should learn to plot several equations on graphing paper and calculate the values to gain insight on the data. This hand-plotting exercise will ensure that the students have a grasp of the fundamental principles of graphing.

I like to get students at the blackboard to graph four plots to offer a pictorial representation of the process before I offer any formal definition. You might use functions such as $f(x) = 2x + 3$, $f(x) = x^2 + 2x + 1$, $f(x) = \sqrt{x-2}$, and $f(x) = \frac{1}{x}$.

In having students make these plots at the blackboard, you want the students doing the work to first tell you what is the domain of the function to be graphed and then make up a table showing ten values of x and f(x). I shall use this recipe to graph the above equations.

Case I: $f(x) = 2x + 3$

You want to get a student to go to the blackboard to generate the table for the function $f(x) = 2x + 3$. First he or she should give you the domain of x which is all real numbers. The student must be able to explain how she or he arrived at the domain of x. This student's explanation helps the other students gain a sense of comfort with this concept.

Table 5.5 (Linear Form) is representative of what you might expect of the graph of the function $f(x) = 2x + 3$. It is important here that students learn to make individual calculations without the aid of graphing calculators that might generate the table. If students work with the data, they gain an appreciate for the graph it will make.

					Table 5.5					
					Linear Form					
x	1	2	3	4	5	6	7	8	9	10
f(x) = 2x + 3	5	7	9	11	13	15	17	19	21	23

The discussion of Table 5.5 is also an opportunity to subtly introduce how to read the slope of a line without overwhelming the students by revealing what you are doing. This equation says for each x value one has to multiply it by 2 and add 3 units more. Simply put, for every move along the run of the x axis one has to rise up the y = f(x) axis by 2 units where the 3 is the initial y value when x = 0.

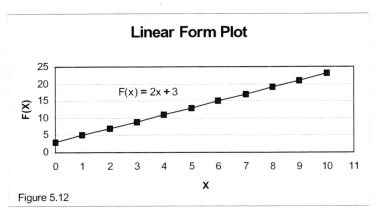

Figure 5.12

Figure 5.12 (Linear Form Plot) is a graph of the equation $f(x) = 2x + 3$. The students can see that it is a straight line. You want to make the point that all linear equations will have similar straight-line graphs although their direction may differ significantly.

Case II : $f(x) = x^2 + 2x + 1$

We will continue our graphing lesson by having a new student go to the blackboard to generate the table for the function $f(x) = x^2 + 2x + 1$. First, you need to have this student give the domain of x, which is $-\infty < x < \infty$ since there are no real numbers that cause undefined situations.

Table 5.6 Second Degree Polynomial								
X		-4	-3	-2	-1	0	1	2	3
$f(x) = x^2 + 2x + 1$		9	4	1	0	1	4	9	16

Table 5.6 (Second Degree Polynomial) is what the student can develop. Here you want to highlight that this table is nonlinear because there is no constant increase or decrease in the f(x) with values of x. You might also suggest to the student at the blackboard that four negative values, zero, and four positive values should be ample data points to gain an appreciation of what the graph looks like. However, in Figure 5.13 (Second Degree Polynomial Plot) we plot 21 points to show a good look at this graph.

Figure 5.13 shows students how to plot the parabola. Students should understand that equations of form $f(x) = ax^2 + bx + c$ give graphs similar to Figure 5.13 although the bowl shape may face down as well as up. It should be pointed out that this graph has a definite minimum value over some region (e.g., $-5 < x < 5$). This comment prepares the students to understand the concept of increasing and decreasing functions over an interval.

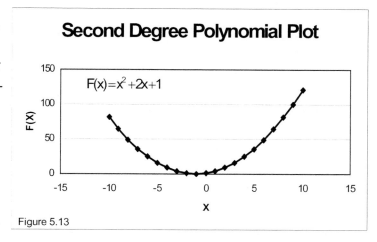

Figure 5.13

Case III: $f(x) = \sqrt{x-2}$

The students are now ready to look at graphing an equation that has a radicand. Call a new student to the blackboard and have her or him define the domain of the function $\sqrt{x-2}$.

You ought get $x \geq 2$, but if that is the only answer given you want to challenge the student to state why this is the domain of x. The student should speak to the need to obtain real solutions in order to make a plot and that is only true if $x - 2 \geq 0$.

Table 5.7 Radical Form										
x	2.00	2.25	2.50	2.75	3.00	4.00	5.00	6.00	7.00	8.00
$f(x) = \sqrt{x-2}$	0.00	0.50	0.71	0.87	1.00	1.41	1.73	2.00	2.24	2.45

The student should now develop Table 5.7 (Radical Form). You might want to point out that this is a nonlinear equation and the students may need a few extra values between 2 and 3 to fully appreciate the curvature in the graph he or she intends to generate. It will enhance the understanding to calculate the f(2.25), f(2.5), and f(2.75) in addition to f(x) where x is the set of integers from 2 to 8 to fill out Table 5.7.

Figure 5.14 (Radical Form Plot) offers the student a pictorial representation of $f(x) = \sqrt{x-2}$.

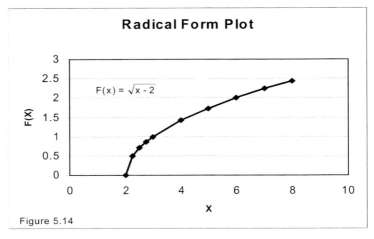

Figure 5.14

There is a significant amount of curvature occurring between x = 2 and x = 3 that tells the student that it may be necessary to use input values that are not even integer increments.

At this point, the student should be getting proficient at using a non-graphing calculator and plotting graphs. These students' feel for the data makes the introduction of more complex items go a lot better because you are building on a foundation in the students' minds versus attempting to teach each new item from fundamental principles. Students should also become hooked on learning, and their desire to pursue more complex material suggests that you do not have to spend a great deal of time motivating them to learn.

Students failing to get enthralled on learning at this point may be too lazy to do their homework assignments and will most likely be lost for the remainder of the semester. Nevertheless, they will feel great peer pressure to catch up because the majority of the class should be working hard to master the material. You might want to offer extended office hours for people who say they want help for various reasons. I find you can often catch up persistent students, but you should not waste your valuable time on people merely looking for ways to avoid hard work. On the other hand, it is important that *you keep in mind* that you have an adult population who also have personal crises that must be resolved on a routine basis.

Case VI: $f(x) = \dfrac{1}{x}$

You want to put a new student to the blackboard. He or she should give the domain of x and then make up Table 5.8 (Basic Rational Function).

First Semester Pre-calculus Under a New Paradigm

Here you want her or him to come up with values for $f(x) = \frac{1}{x}$ on both sides of zero. It is very instructive here to have the student calculate several values close to zero that will make learning the concept of the asymptote become obvious.

Table 5.8
Basic Rational Function

x	-3.0	-2.0	-1.0	-0.8	-0.5	-0.3	-0.1	0.1	0.3	0.5	1.0	2.0	3.0
$f(x)=\frac{1}{x}$	-0.3	-0.5	-1.0	-1.3	-2.0	-4.0	-10.0	10.0	4.0	2.0	1.0	0.5	0.3

In plotting the data in Table 5.8, the students see that the curve approaches a line with the equation x = 0. Figure 5.15 (Reciprocal Plot) shows the points approaching the line x = 0 from both directions. You might call the students' attention to the concept of the graph approaching the x = 0 line and you label the line the asymptote. Emphasize that there is a discontinuity in this graph and it occurs precisely at the asymptotic point x = 0.

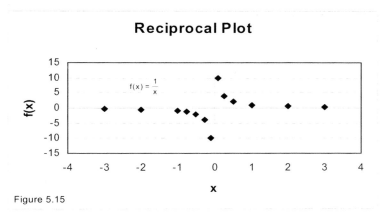

Figure 5.15

Vertical Line Test for a Function

Barnett and Ziegler offer us a theorem for determining whether or not we have a function (Barnett & Ziegler, p. 200).

"An equation defines a function if each vertical line in the rectangular coordinate system passes through at most one point on the graph of the functions. If any vertical line passes through two or more points on the graph of an equation, then the equation does not define a function." This statement is given in a classic mathematics mind-set. There is no doubt here that it is a theorem and not a definition.

Larson, et al., offer a more user-friendly version of this theorem. "Vertical Line Test for Functions, A set of points in a coordinate plane is the graph of y as a function of x if and only if no vertical line intersects the graph at more than one point" (Larson, Roland & Hostetler, p. 117).

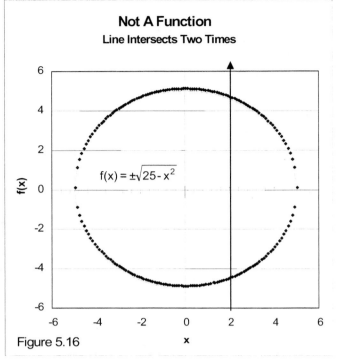

Figure 5.16

You might parse this theorem and talk the students through it.

If and only if no vertical line intersects the graph at more than one point: Here a picture is worth a thousand words, so you want to make graphs. Two functions that are apropos are $F(x) = \pm\sqrt{25 - x^2}$ and $F(x) = x^3 + 2x^2 + 3x + 4$.

Figure 5.16 (Not A Function) shows a line intersecting the graph of $F(x) = \pm\sqrt{25 - x^2}$ at two points. This intersection at two points says that F(x) is not a function because it violates the theorem. I like to stress that *there can be no deviations from definitions and theorems in mathematics*. Either your situation meets the definition or theorem in totality or it fails.

You want to talk through the graph speaking to the fact that each x value generates two different F(x) values. Furthermore, there is no unique F(x) solution for a given x.

Here is an opportunity to remind the students about the real number line where they learned about the uniqueness of numbers. On the real number line there is a one-to-one correspondence between the real numbers and the positions on the line.

First Semester Pre-calculus Under a New Paradigm

If the students examine Figure 5.16, they see that there is no one-to-one correspondence. This reintroduction of the one-to-one correspondence concept reinforces the students' understanding of this concept and prepares them to be ready to handle inverse functions.

An examination of Figure 5.17 (A Function Line Intercepts At One Point) shows that a vertical line will only cross the plot of the curve once. This curve meets the vertical test; therefore, F(x) is a function.

When the students look at the curve in Figure 5.17, be sure they notice that it crosses the x axis at least once. You want to encourage the students to see that where the curve crosses the x axis that the F(x) = 0. *It is very important that the students realize that F(x) = 0 all along the x axis. Even though you might think that is obvious, you may find that many students struggle with that point, especially if this is their first study of algebra.*

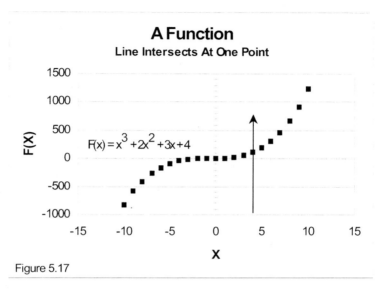

Figure 5.17

Before we continue our discussion of functions, it is important to talk to the students a bit about why they are using their valuable time to study this subject. In multinational corporations, one often wants to use mathematical modeling in many situations versus spending large sums of money if he or she can avoid making expensive prototype products. If one can derive or find a formula that describes his or her operation, it allows them to plug in values and have an idea of what they might expect at a particular value.

If we look at Figure 5.14 and Figure 5.15, the prediction process is straightforward. The student puts in a value for x and gets a unique value for F(x). One can feel reasonably certain that he or she has a value that is representative of his or her process.

Chapter 5

However, in Figure 5.16 students are in the precarious position of trying to decide which F(x) value to use. This uncertainty means that the students' mathematical model did not give the definitive information needed to make a decision on the potential merits of the product under contemplation. This suggests that the students will need to go through the costly effort of building various versions of the prospective product. It means students must question their mathematical model because the equation is not representative of the process they wish to mimic. In either case, the student does not have a response to the dictates of his or her management's goals or objectives.

You can make this function concept personal in the students' eyes in assigning a test problem or having a class discussion where the students give two hypothetical pay amounts and learn to figure out how to estimate their raises in future years. Perhaps students pick 30,000 dollars pay in 2000 and 36,000 dollars in 2003. What would the person expect to earn for the year 2006?

$$m = \frac{36,000 - 30,000}{2003 - 2000} = \frac{6000}{3} = 2,000$$

Although we have not given a detailed discussion of the slope, the students can model this problem by using their limited understanding of slope and the point slope equation. Student readily calculate the slope.

The point slope equation tells the students that Y - 30,000 = 2,000(X - 2,000) or they can chose the equation Y - 36,000 = 2,000(X - 2003).

Have a student work through these calculations at the blackboard. He or she should devise a reply similar to the following:

Y - 30,000 = 2,000(X - 2000)

Y = 2,000 X - 2,000*2000 + 30,000

Y = 2,000X - 3,970,000

Y = 2,000*2006 - 3,970,000 For calculating the student's pay in the year 2006.

Y = 42,000

If students want the pay in the year 2006, they need to substitute this value for the independent variable X. The issue is that once the students calculated the slope of 2,000 they needed to first ask themselves what does this number reveal. This gives you an opportunity to have everyone gain confidence in the fact that the slope of a straight line is the same no matter what two points are chosen.

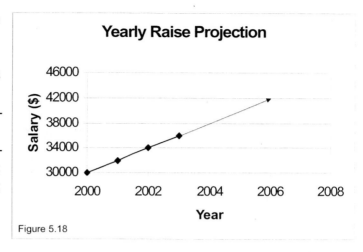

Figure 5.18

This understanding of slope will allow the students to use this concept to predict future pay raises.

A slope of 2,000 dollars here tells the students that each year their pay will increase by two thousand dollars. Another way to look at this concept is for each unit you increase on the graph *run* axis you must go up the graph *rise* axis by 2,000 units (see Figure 5.18). Hence, the students do not need to go through the above calculation; they can merely say each year it is rising by 2,000 dollars, so three years at that rate equals 6,000 dollars. Add that amount to the 36,000 for the year 2003. This commonsense calculation gives 42,000 dollars.

The ideal of having a clearly defined function as a straight-line equation leaves no doubt as to what the pay raise will be. Students see how one can use their mathematics to make predictions.

$$r = \frac{\frac{24(NM-P)}{N}}{P + \frac{NM}{12}}$$

Since we are on the subject of using the mathematics to affect one's personal life, this is a good place to offer the students an equation that Larson, Hostetler, and Edwards say gives the approximate annual interest rate "r" of a monthly installment loan (Larson, Roland & Hostetler, p. 48).

In the above equation N is the total number of payments, M is the monthly payment, and P is the amount financed. This equation translates the advertised loan rate into a yearly loan rate that people will actually pay on a loan. It is a rude awakening for the students when you have them work through the difference in actual and advertised rates.

Increasing and Decreasing Functions

If you plow into the textbook definitions of increasing and decreasing functions, you may find that you lose many people. However, if you recount some information that students already know, then these concepts are not that difficult to comprehend. First you want to remind the students that you are going to use the definition of $y = F(x)$ and plot a graph. Instead of having the *rise* axis labeled y, you will label it the $F(x)$. Let us use three different functions to make our case to the students so they will not wonder if this technique is only good for one type of problem.

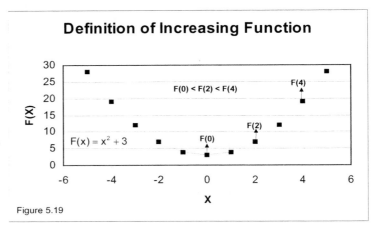

Figure 5.19

We shall use $F(x) = -2x + 3$, $F(x) = x^2 + 3$ and $F(x) = 3$ for our discussion. The first step is to allow the students to see what happens to $F(x)$ as the "x" values increase. We need to define a region over which we are interested in examining our function. Then we want to determine whether $F(x)$ is increasing, decreasing or constant in the interval of interest.

Let us start our discussion by having the students examine Figure 5.19 (Definition of Increasing Function) where we plotted the function $F(x) = x^2 + 3$. If we restrict our discussion in the domain of x to the interval $0 \le x \le 5$, then the students readily see that $F(0) < F(2) < F(4)$. $F(x)$ is increasing if $F(x_0) < F(x_2) < F(x_4)$ when $x_0 < x_2 < x_4$.

Thus if you go over one of the classic definitions of an increasing function such as, "Let I be an interval in the domain of a function f. Then f is increasing on I if f(b) > f(a) whenever b > a in I" (Barnett & Ziegler, p. 213).

You want to talk through this definition. The interval over which we are interested "I" is the above expression $0 \le x \le 5$. The students can see as the x values ascend from 0 to 5 that $F(x)$ gets larger and larger; e.g.,

$F(5) > F(4) > F(3) > F(2) > F(1) > F(0)$. This says that this function in the interval we defined is an increasing function because it meets the definition of an increasing function.

If we continue with our definitions, "f is a decreasing function on I if f(b) < f(a) whenever b > a in I."

Figure 5.20 (Decreasing Function) gives a pictorial representation of the decreasing function. The students can see that as the x values grow larger and larger the F(x) values are growing smaller and smaller. This condition meets the definition because $F(x_1) > F(x_2)$ whenever $x_2 > x_1$ over the x domain which is all real numbers.

It is good to have a student plot Figure 5.20 at the blackboard. Nonetheless, the hidden agenda here is to encourage students to read the definitions and learn to apply them. You want the students to pursue understanding whilst moving these students as far as possible away from the practice of memorizing without comprehending what is really happening.

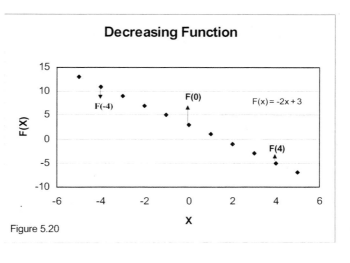

Figure 5.20

Heretofore, we have looked at increasing and decreasing functions. Now let us turn our attention to the constant function. Some students might struggle with the concept of a constant function because they have become accustomed to plotting the independent variable "x" versus the dependent variable F(x) where they find different F(x) results for each x value. Having a student make the graph of F(x) = 3 is very instructive here, although we will make a table of values that will help the students to learn to think numerically.

Table 5.9 Constant Equation										
X		-5	-4	-3	-2	-1	0	1	2	3	5
F(x) = 3		3	3	3	3	3	3	3	3	3	3

Table 5.9 (Constant Equation) suggests that whatever value for the independent variable we choose, F(x) is always the constant value 3. This table allows the students to see a numerical illustration that meets the definition "f is constant on I if f(a) = f(b) for all a and b in I." Since I is all real numbers and the students see that $F(x_1) = F(x_2)$ for x_1, Table 5.9 meets the definition of a constant function.

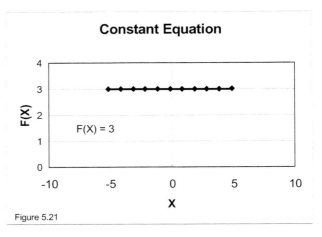

Figure 5.21

Although we want to encourage students to learn to think numerically, it is still instructive to see a pictorial representation of Table 5.9, especially for students who may have difficulty visualizing this numerical representation. Figure 5.21 (Constant Equation) offers a pictorial representation of Table 5.9. You want to further point out to the students that clearly F(x) maintains the same value regardless of what the x may be. This makes F(x) a constant because this concept means the lack of change.

Relative Maximum and Relative Minimum

Discussions of relative minimum and relative maximum have more relevance when you couch them in the need for going through these exercises. The students need to understand that equations and graphs created heretofore will someday describe various activities. Some activities can impact a process in their plants and others touch on the amount of money they may receive or have to pay out because of their personal or business activities. Surely students (managers and professionals) will want to know when they are at the peak on a process or on the other hand when it bottoms out and they can no longer look forward to taking out some expense component. First let us discuss the relative maximum definition. "A function value f(a) is called a relative maximum of *f* if there exists an interval (x_1, x_2) that contains *a* such that $x_1 < x < x_2$ implies $f(a) \geq f(x)$" (Larson, Roland & Hostetler, p. 119).

What the students want to feel comfortable understanding here is that this definition says in layman's terms that in some interval there is a point (we choose to call this point "a") on the independent variable axis that gives an f(a) that is bigger than every other f(x) value one can generate by choosing any other x value.

Figure 5.22 (Relative Maximum) offers a graph with two relative maximums that fit the definition for relative maximum. If we choose the interval $-3 < x < 0$, the students see that a relative maximum occurs at approximately $x = -1.7$ because in this interval $F(-1.7) > F(x)$. A similar situation occurs in the interval $0 < x < 3$ where the students see another relative maximum occurs at approximately 1.8 because $F(1.8) > F(x)$.

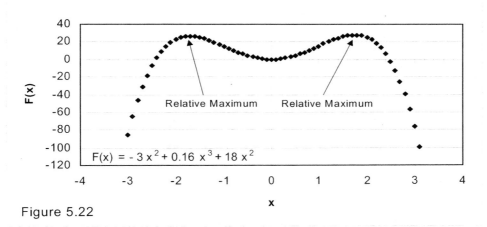

Figure 5.22

Let us now discuss the relative minimum definition. "A function value f(a) is called a relative minimum of f if there exists an interval (x_1, x_2) that contains a such that $x_1 < x < x_2$ implies $f(a) \leq f(x)$."

Figure 5.23 (Relative Minimum) offers a graph with two relative minimum points that fit the definition for a relative minimum. If we choose the interval $-2.25 < x < 0$, the students see that a relative minimum occurs at approximately $x = -1.5$ because in this interval $F(-1.5) < F(x)$. A similar situation occurs in the interval $0 < x < 2.5$ where the students see another relative minimum occurs at approximately 1.75 because $F(1.75) < F(x)$.

It is important that you remind the students that finding these solutions graphically and numerically were excellent methods prior to the invention of graphing calculators and graphing routines in spreadsheets because there was a need to be able to trace a curve without making numerous calculations with the potential for errors in manual calculations or reading a slide rule.

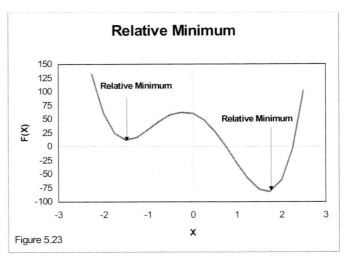
Figure 5.23

Before we go we need to see a function with an *absolute minimum*. We want to find a point c such that $f(c) \leq f(x)$ for all x in the domain. Figure 5.24 (Absolute Minimum) offers us a plot of the expression $f(x) = 3x^4 - 16x^3 + 18x^2$ where we see that the absolute minimum occurs as x = 3 and f(3) = -27. Using a similar argument, we define the absolute maximum as being at the point c such that $f(c) \geq f(x)$ for all x in the domain.

Even and Odd Functions

Barnett and Ziegler offer a definition to cover even and odd functions. "A function is called an even function if its graph is symmetric with respect to the vertical axis and an odd function if its graph is symmetric with respect to the

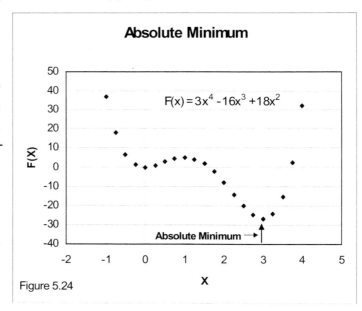
Figure 5.24

origin" (Barnett & Ziegler, p. 228). They also offer a theorem to make the even and odd functions concepts useful in their application. "If f(-x) = f(x) the *f* is an even function. If f(-x) = -f(x), the *f* is an odd function."

If you explain this definition, you want to have pictorial representations to guide the students through the concept of symmetry.

First consider an even function. The definition says it should be symmetric about the vertical axis. We need to define symmetry in layman's terms. That is, you see the same shape graph going the opposite direction on the other side of the vertical axis. Figure 5.25 (Even Function) is a pictorial representation of an even function.

However, some students may have difficulty grasping the even concept from the theorem.

They do not fully comprehend that f(-x) = f(x) in an even function. Problems such as $f(x) = 6x^{-6}$ will be labeled an odd function because students missed the fact that the minus sign in the exponent merely means that this variable shifts to the denominator. We can rewrite the above expression as $f(x) = \frac{6}{x^6}$. In this new form the students can see it is an even function because f(-x) = f(x).

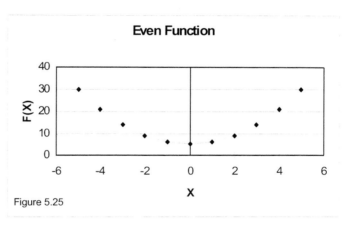

Figure 5.25

Students have few problems with even functions such as $f(x) = 2x^2$. Here they can see that one can substitute -x for x and get f(x).

Let us turn our attention to odd

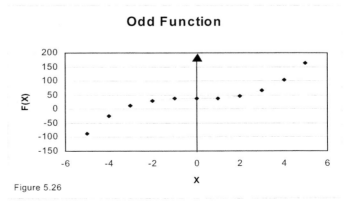

Figure 5.26

functions such as $f(x) = x^3 + 3x$. Figure 5.26 (Odd Function) offers a pictorial representation of this odd function. The students see that it is symmetric across the origin. Also it is apparent in Figure 5.26 that $f(-x) = -f(x)$ which follows what the theorem indicates should be the case.

There are many functions that do not satisfy the definition for an even or an odd function. These function are neither even nor odd. For example, $F(x) = 3x^3 + 2x^2 + 5x - 10$ is neither even nor odd since replacing x by -x results in neither $F(-x) = F(x)$ nor $F(-x) = -F(x)$. Therefore, these types of functions are neither even nor odd. Figure 5.27 (Neither Even nor Odd Function) is an example of an expression that is neither even nor odd. You will notice that it appears symmetric in respect to the point $x = 1$ versus the origin.

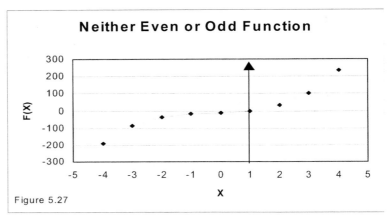

Figure 5.27

Absolute Value Expressions

In discussing graphing an absolute value expression, you want to call upon the students' background by trying to determine whether an expression such as $f(x) = \pm\sqrt{x^2 - 9}$ is a function. This expression has two different values for f(x) being generated from the same value of x. With that thought in mind and returning to the definition of the absolute value of a, $|a| = a$ when $a > 0$, and $-a$ when $a < 0$. The students can fully appreciate the graph of $F(x) = |x|$ (see Figure 5.28 Absolute Value).

This absolute value of x equation does satisfy the definition of an even function because substituting $F(-x) = F(x)$. Furthermore, the absolute value expression is symmetrical around the vertical line.

It is helpful to go through the absolute value expression whenever the opportunity avails itself because this is a concept that may take a few exposures before it becomes a common expression in the students' psyche.

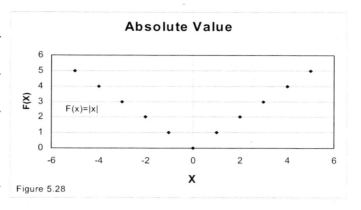
Figure 5.28

The families of expressions we have studied heretofore have included expressions such as absolute value, constant, square root, squaring functions, cubing functions, and so on. What you want the students to understand is that they now have an idea of what the graphs of these equations look like.

Students also need reminding that when they make graphs for expressions such as F(x) = 2x² + 3x + 10, that when they let x = 5 it generates the F(5) = 2(5)² + 3*5 + 10. Students' understanding of this substitution concept is critical to their comprehending business calculus in the future because this is a concept with which some business calculus students struggle.

Vertical and Horizontal Shifts

If your students have manually graphed many equations, they can usually see what happens when a constant is added to the independent variable in an expression raised to a power or it is added or subtracted. Plotting graphs as Figure 5.29

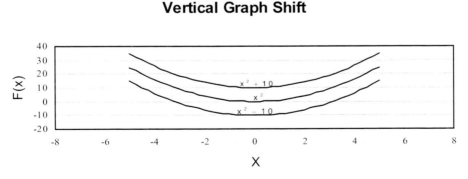

Figure 5.29

(Vertical Graph Shift) where you have $F(x) = x^2$, $F(x) = x^2 - 10$, and $F(x) = x^2 + 10$, you can ask the students what will happen in each case and they should be able to provide the proper response to your question. The key learning in Figure 5.29 is we have a vertical shift.

In Figure 5.30 (Horizontal Graph Shift), we include the shift with the independent variable. Here we will move on either side of the $x = 0$ line. Let the students now plot the equations $F(x) = (x-10)^2$, $F(x) = x^2$ and $F(x) = (x+10)^2$ that make up Figure 5.30. Most students should be able to say what will happen without making Figure 5.30, but for instructive purposes we shall include this figure. The key learning in Figure 5.30 is we have a horizontal shift that occurs in our graph. The positive and negative values added to the independent variable cause curves that flow in opposite directions around the vertical axis.

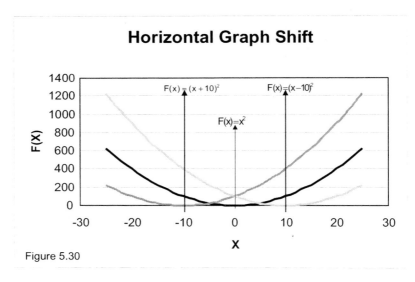

Figure 5.30

Reflections

We can now turn our attention to the concept of multiplying our independent variable by both positive and negative constants. First let us look at reflections where we multiply an expression by minus one and simultaneously plot the new graph on the same coordinate system. Consider the expression $F(x) = 2x^2$. Figure 5.31 (Reflection Expression) shows that the students can generate the reflection of F(x) by setting F(x) = -F(x). The students see when you plot the $F(x) = -2x^2$ on the same coordinate system as our original $F(x) = 2x^2$ they get a reflection of the graph across the x axis.

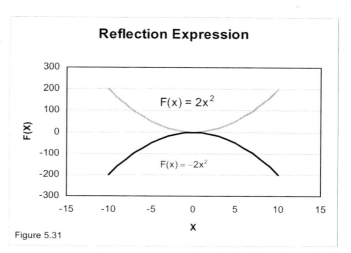

Figure 5.31

Expansion and Contraction

We want to keep in mind that our goal is not merely to have students blindly plot equations, but also to have our students gain a feel for what various equations will look like from merely examining their generating equations. Thus, we will end our discussion of graphing with an examination of what happens when we have $y = c f(x)$ where $c > 1$ or $0 < c < 1$.

Before you have a student draw any graphs, it is very instructive to have a student tell what will happen when you multiply f(x) by 2. Ask her or him to describe precisely what they will do to generate graphs with both f(x) and 2f(x) on the same coordinate system using the same x values. You should hear something to the effect that they will put in x, generate the value of f(x), and then multiply the first $f_1(x)$ by 2 to generate $f_2(x)$.

Give the student expressions such as $f(x) = x^2$ and $f(x) = 4x^2$ and have her or him generate a table of values for the domain $-5 \leq x \leq 5$. Table 5.10 is the generating values for the student's graph that you will have her or him to make. You want the student to tell you what this table says, but this is another opportunity to help people learn to think numerically. You might stress

Table 5.10											
Contraction Equation											
x	-5	-4	-3	-2	-1	0	1	2	3	4	5
$F(x) = x^2$	25	16	9	4	1	0	1	4	9	16	25
$F(x) = 4x^2$	100	64	36	16	4	0	4	16	36	64	100

that *tables tell stories* and the students' management in industry will expect them to give a narrative describing what the numbers reveal. Do not allow the students to skip this exercise because you will encumber their ability to learn to think numerically whilst still in college and also hamper your ability to help your students to etch pictorial representations of data in their psyche. By the time students finish pre-calculus-I, they should have a feel for what curves may look like from their generating equations and also a hint from the data.

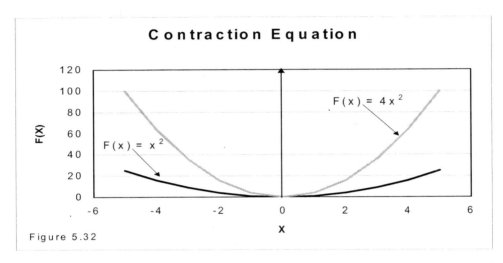

Figure 5.32

Figure 5.32 (Contraction Equation) is a pictorial representation of Table 5.10. It is readily seen that multiplying F(x) by 4 caused the new curve to rise quickly with each x value versus the initial expression. This *multiplying F(x) by a constant greater that one causes a constricting* on the new curve relative to the original one; therefore, our students see this action as contraction of the curve.

On the other hand, you want to challenge the students to think about what happens when they multiply F(x) by a constant whose value is $0 < c < 1$. Have a student describe what happens in this situation prior to making a table or graph. Encourage the student to look at the problem in its most fundamental form. That is, what happens to each new F(x) value when the old value is multiplied by a number that is less than one and greater than zero? Our student should come to the conclusion that $|F(x)| > |cF(x)|$ when $0 < c < 1$. This conclusion tells the students that for each value of x the absolute value of cF(x) will be a smaller value on the vertical axis. Hence, the curve generated will have a wider horizontal opening for a given x value.

Figure 5.33 (Expansion Equation) offers three graphs that combine our discussion heretofore on expansion and contraction. The students see pictorial representations of the expressions $F(x) = x^2 + 30$, $F(x) = 2x^2 + 30$, and $F(x) = \frac{1}{2}x^2 + 30$. The students see expressions where $0 < c < 1$, $c = 1$, and $c > 1$. Clearly the graphs in Figure 5.33 are wider for a given x value as the c value decreases from $c > 1$ to $c = 1$ to $0 < c < 1$.

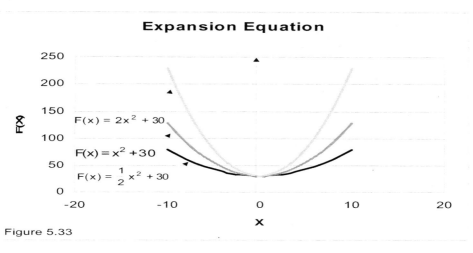

Figure 5.33

It is worthwhile at this point to have several students at a time come to the blackboard and you provide them with some problems to be worked out. You want to see that every student has a feel for the concepts you have spent a great deal of time discussing. Look for people's ability to make a table and then graph the expression that you give them. *Do not permit graphing calculators whilst students are at the blackboard*; however, standard calculators are okay. Graphing calculators take away the opportunity for students to work with the data suggesting that they will get less of an intuitive feel for the data with these instruments.

Make sure that you give problems that cover all of the areas you discussed here. Also you will want the degree of difficulty to increase as different people go to the blackboard because the students observing should be learning from the people doing their work in front of the class. This technique also helps to keep students paying attention to what is going on.

Students who do not grasp these graphing and numerical data representation techniques immediately should remain at the blackboard until they garner an understanding. If time does not permit the student to obtain an understanding of these points during class time, you should set up offices hours or help the student find a tutor.

It is good to develop a table for students to use when thinking about graphing functions. Barnett and Ziegler offer a summary of graphing assistants entitled, "Graphing Aids for Functions" (Barnett & Ziegler, p. 234), found in Table 5.11, that summarizes our learning heretofore. You want to strongly encourage students to understand Table 5.11 because it is an excellent aid in fostering their learning to think through problems before setting out to come up with solutions.

After you have assured yourself that the students understand how to manually make graphs, you can then allow them to use graphing calculators or graphing routines on their personal computers. The *danger of allowing the use of graphing calculators too soon is that you teach calculator and not mathematics.* Students may never really learn to plot an expression from fundamental principles and gain the insights that accompany that exercise.

I worry that many students who overwhelmingly use graphing calculators in the early phases of learning the fundamental principles of pre-calculus will fall victim to a mere perturbation of their present enchantment with memorizing mathematical formulas without understanding these expressions. Students often share how they memorized equations to pass a course but they never really understood what occurred throughout the semester.

I make a concerted effort to reprogram my students away from their penchant for memorization to one of seeking to understand each new mathematical concept. I often proclaim, *"If you understand, you can create. If you don't understand, you can only do what someone else tells you to do."*

Sum, Difference, Product, and Quotient of Functions

The students should know that they have made significant progress to reach this point in their study. You can start making subtle gestures that the difficult task is behind the students for now they have an excellent foundation on which to build the concepts of pre-calculus, finite mathematics, and business calculus. Your students should come to understand that for the remainder of the semester, you will offer techniques for handling various situations in the use of equations.

Table 5.11
Graphing Aids for Functions

Symmetry Test

$f(-x) = f(x)$	Graph of $y = f(x)$ is symmetric with respect to the vertical axis
$f(-x) = -f(x)$	Graph of $y = f(x)$ is symmetric with respect to the origin

Vertical Translation

$y = f(x) + k,\ k > 0$	Shifts the graph of $y = f(x)$ upward k units
$y = f(x) - k,\ k > 0$	Shifts the graph of $y = f(x)$ downward k units

Horizontal Translation

$y = f(x - h),\ h > 0$	Shifts the graph of $y = f(x)$ to the right h units
$y = f(x + h),\ h > 0$	Shifts the graph of $y = f(x)$ to the left h units

Reflection

$y = -f(x)$	Reflects the graph of $y = f(x)$ in the x axis

Contraction

$y = Cf(x),\ C > 1$	Contracts the graph $y = f(x)$ by multiplying each coordinate value by C.

Expansion

$y = Cf(x),\ 0 < C < 1$	Expands the graph $y = f(x)$ by multiplying each coordinate value by C.

Chapter 5

As you talk through the sum, difference, product, and quotient, it *brings confidence in the students to see you pull various equations from your mind* as you present examples to explain your points. You are going to be making a de facto presentation and the students see that you can think on your feet. Thus, you encourage your students, by example, to think on their feet. I am a firm believer if the students do not feel you have a good mastery of the material, you lose them for the term, so you want to seek oblique opportunities to demonstrate your knowledge and not come over as a grandstander, which will only turn off your students if they conclude that you are a condescending professor.

Consider two functions f and g with overlapping domains and let us define some operations with them.

Sum

$$(f + g)(x) = f(x) + g(x)$$

If $f(x) = 3x^2 + 4x$ and $g(x) = 7x^3 + 9x^2 + 7$ find the $(f + g)(x)$.

$$f(x) + g(x) = 7x^3 + 12x^2 + 4x + 7$$

Difference

$$(f - g)(x) = f(x) - g(x)$$

If $f(x) = 9x^3 - 6x^2 - 3\sqrt[4]{x^3}$ and $g(x) = -9x^3 - 6x^2 + 2x^{0.75}$ find the $(f - g)(x)$.

$$f(x) - g(x) = 18x^3 - 5(x^{0.75})$$

Note: $x^{\frac{p}{q}} = \sqrt[q]{x^p} = (\sqrt[q]{x})^p$ This means $\sqrt[4]{x^3} = x^{\frac{3}{4}} = x^{0.75}$

Product

$$(fg)(x) = f(x)g(x)$$

If the $f(x) = 2x - 3$ and the $g(x) = 5x^2 + 9$, find the $(fg)(x) = f(x)*g(x)$.

$$(2x-3)(5x^2 + 9) = 10x^3 - 15x^2 + 18x - 27$$

Quotient

$$(f/g)(x) = (f(x)/g(x) \text{ if } g(x) \neq 0.$$

If $f(x) = 4x^2 - 10$ and $g(x) = x^2 - 9$, find the $f(x)/g(x)$.

The first step here is to pay attention to the restriction that $g(x) \neq 0$ which tells the students that the domain of x is all x except that $x \neq \pm 3$.

$$\frac{f(x)}{g(x)} = \frac{4x^2 - 10}{x^2 - 3}$$

Composite Function

The composite function is an area you want to stress highly. The lack of understanding of this function is a death-knell when you try to define the derivative in calculus.

You might consider starting this discussion by having the students recall that they have already heard of a composite number. The composite number is one that is made up of multiplying prime numbers together. The number 18 is composite because it is made up of 2*3*3.

Following this composite number mind-set where one number is made up of other numbers, let us do something similar with a function where the variable can have different values depending on what we want to substitute for it to generate our new function. The key difference in our analogy with using prime numbers as building blocks to generate a composite number is there is only one specific set of prime numbers to generate a composite number but there are numerous values we substitute in our composite function as long as they do not render it undefined.

Barnett and Ziegler offer the definition of a composite function as, "Given functions f and g, then $f \circ g$ is called their composite and is defined by the equation $(f \circ g)(x) = f[g(x)]$. The domain of $f \circ g$ is the set of all real numbers x in the domain of g where g(x) is in the domain of f" (Barnett & Ziegler, p. 259-261).

You want to parse this definition and go through each section to get the students comfortable with what it says. Let us examine a composite function example from Barnett & Ziegler where you can see the definition parsing through following the problem solving steps.

"Find $(f \circ g)(x)$ and its domain for

" We begin by stating the domains of f and g, a good practice in any composition problem:

$$\text{Domain } f: -2 \leq x \leq 2 \text{ or } [-2, 2]$$
$$\text{Domain } g: x \leq 3 \text{ or } (-\infty, 3)$$

Next we find the composition:

$$(f \circ g)(x) = f(g(x)) = f(\sqrt{3-x})$$

$$= \sqrt{4 - (\sqrt{3-x})^2}$$

$$\left(\sqrt{t}\right)^2 = t \quad t \geq 0$$

$$= \sqrt{1+x}$$

"Even though $\sqrt{1+x}$ is defined for all $x \geq -1$, we must restrict the domain of   to those values that also are in the domain of g. Thus, Domain $f \circ g: x \geq -1$ and $x \leq 3$ or $[-1, 3]$."

The key issues are the students must first find the domain of x for f(x) to find the region of concern. Using the defined region for f(x), the student can find the subset for the g(x) to narrow down the possible values in our discussion. Knowing the set of values of x for the g(x), the students now find the set of values of x for the f[g(x)] from those values.

One-to-One Function

When we defined the function, we used the vertical line test to show that it only touched the graph of our expression once in the domain of x. This same mind-set is captured in a definition of a one-to-one function *f* for values on its domain which says $f(x_1) = f(x_2)$ implies that $x_1 = x_2$.

If you have the students recall an earlier discussion on the real number line where they learned that there is a one-to-one relationship with positions on this line and the real numbers, then they have a concept of one-to-one in mind. A key point to stress is the *uniqueness of numbers*. There is only one space for the number one on the real number line. Anything occupying the number one location must be number one.

If we now turn our attention to the discussion of inverse functions, then we can make the case to the students that a function R has the inverse R^{-1} if and only if R is one-to-one. Without having a one-to-one correspondence, as is not the case in the expression $x = y^2$, you have no idea of which value of y to use when you consider that $y = \pm\sqrt{x}$. The key concern is that the same value of x generates two different values of y.

Inverse Function

Since the students are probably wondering what exactly is an inverse function, it is now time we define it. You want to call upon their understanding of composite functions and one-to-one correspondence to present your definition. Let us now define the inverse function as $f(g(x)) = x$ for every x in the domain of g and the $g(f(x)) = x$ for every x in the domain of f. The function g is denoted by f^{-1}. This tells us that $f(f^{-1}(x)) = x$ and the $f^{-1}(f(x)) = x$. That is, the domain of f must equal the range of f^{-1}, and the range of f must equal the domain of the f^{-1}.

A good way to see this is to first generate an inverse function and then test it against the definition. We note that the last portion of the definition says that domain of a given function becomes the range of its inverse. This allows us to change the values x and y and solve the new function for x. Take the function $y = \sqrt[3]{x+6}$ and work through it with the students to find an inverse. The domain of x in this function is all real numbers which will be the range of its inverse function.

$$x = \sqrt[3]{y+6}$$

$$x^3 = y + 6$$

$$x^3 - 6 = y$$

We are now ready to check both of these equations to see if they fit the definition.

Case I: $y = \sqrt[3]{x+6}$

$$y = \sqrt[3]{x^3 - 6} + 6$$

$$y = x$$

Case II: $y = x^3 - 6$

$$y = (\sqrt[3]{x+6})^3 - 6$$

$$y = x$$

As can be seen in the above, the expression $f = y = \sqrt[3]{x+6}$ has $f^{-1} = y = x^3 - 6$ as its inverse function.

EQUATION

An *equation* is a statement that two algebraic expressions are equal to each other ($y = 2x + 9$). However, the *solutions* to an equation are those values that make the statement true.

A linear equation in one variable x is an equation that can be written in the standard form $ax + c = 0$ where a and c are real numbers and $a \neq 0$. It is helpful if you have a student solve this equation for x to show the class the importance of being able to do a derivation to solve a family of problems. I make the pitch here that arithmetic helps you to solve only one problem and mathematics helps you solve a family of problems once you have made a derivation.

$$ax + c = 0$$

$$ax + c - c = -c$$

$$\frac{ax}{a} = \frac{-c}{a}$$

First Semester Pre-calculus Under a New Paradigm

$$x = \frac{-c}{a}$$

The students now have a unique solution for this linear equation. They also have an approach for arriving at this unique solution. You want to get the students to do several problems using this concept to solve problems of this linear form.

Problem I - Find the value of x.

$$\frac{x}{5} + \frac{3x}{7} = 9$$

7x + 15x = 315 after multiplying both sides by 35.

22x = 315

$$x = \frac{315}{22}$$

You want to give problems that do not always work out to a nice even solution because in the real world students may not find nifty answers to real problems. A second problem in this mind-set is:

$$\frac{1}{x-3} = \frac{3}{x+3} - \frac{5x}{x^2-9}$$

x + 3 = 3(x - 3) - 5x when original equation is multiplied by (x + 3)(x - 3)

x - 3x + 5x = -9 -3 and x = -4

WORD PROBLEMS

We have not stressed word problems to this point. It is now time to turn our attention to them. *Word problems test the students' ability to read.* At this point in the semester students should be beyond merely calling words to understanding what they say in context. Students should have their confidence built up to speak before their class.

You want to build on this student self-confidence. First, teach the students how to make a picture whenever possible and write down the facts in broaching word problems. Students should be able to affirmatively answer the question, do I have sufficient data to respond to the query before me?

Although students have some self-confidence, you ought to expect some displays of consternation on their part when you change your emphasis to significantly elevating the importance of understanding how to solve word problems. This is, you must encourage your students to adopt tenacity as a key element in their learning.

It is good to remind people that *learning requires stick-to-it-tiveness,* and it may not come in a continuum. Perhaps it occurs more in quantum steps. This means that students will have to stick to doing increasingly difficult word problems until attacking these problems becomes a normal part of their approach to solving problems.

Business students understand when you tell them that requests will often come in the form of letters or memos defining a problem that they are asked to solve. This suggests that understanding how to solve word problems is imperative for their future success in the business world. Once you establish this need to understand word problems, you should see the students' apprehension start to dissipate when they overcome their initial fear of not being able to read well enough to understand what is being asked.

The team approach to learning to solve word problems helps to reduce the students' consternation quickly. Create teams to solve a particular problem and this group must elect a spokesperson to present their effort to the class. After doing several problems in this manner over two or three class meetings you can then start to draft individuals to work at the blackboard. These individuals may display a minor case of the butterflies but when they start working problems they usually overcome any hesitation due to *reading inertia.*

You can further reduce the students' stress by allowing people to work on their assigned problem at their desk prior to being required to put their problem and results on the blackboard. However, *students should not be allowed to take their notes to the blackboard when solving their word problem.* One of the hidden agendas is to teach students how to think on their feet and recover if things do not come out exactly as expected because of an error in their calculations.

There are some basic formulas and geometric and business understandings that students must have to do many word problems. Students may ask for a review of key equations. It is important that you not merely show the students a group of equations. You also need to help them understand the underpinning concepts in these expressions.

Square: The square (Figure 5.34) is a body that has the same measure length on all of its sides. We define the area (A) of a square by multiplying its length s by its width s; so $A = s * s = s^2$. We also define the perimeter (P) by measuring the length of all of the sides of the square; thus, $P = 4s$.

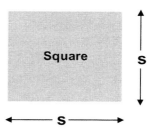

Figure 5.34

The concept of a square has little meaning until you put units to each side. If you have the students visualize a foot, yard, meter, inch, centimeter, mile, or so on, then discuss a square as the material, the land, or whatever that measures either one foot by one foot by one foot by one foot or the same pattern for meters, inches, and so on. Thus, the area becomes one square foot because we add the power on units as we do for any algebraic expression when we multiply them.

This square becomes a tool for determining how big other surfaces are because you can calculate how many of these squares it takes to cover the areas you have in mind. We will discuss this point further when we discuss rectangles.

You can extend our square discussion to talk about its perimeter. If the students wish to determine how many inches or feet or meters there are around this square, they can accomplish this task by adding the lengths of each side around the square. That is its *perimeter* (4s).

Rectangle: The rectangle (Figure 5.35) is a body with four sides whose angles are right angles and the opposite sides have the same measure. This tells us that its perimeter (P) is two times its length plus two times its width. We express perimeter algebraically as $P = 2l + 2w$.

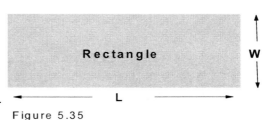

Figure 5.35

We define the area (A) of the rectangle as A = L*W. You want to tell the students that a square is just a special kind of rectangle where all of the sides have the same measure.

Since the rectangle is the general case, students can ask themselves how many of the squares does it take to cover the area in a rectangle? It is the answer to that question that makes the area concept live in the students' minds. Just pose the question, "If you have a living room that measures 12-feet by 14-feet, how many one-foot-square tiles will it take to cover this room?"

You want the students to understand that they must first calculate the area of the living room and then divide that area by the area of a single tile to know the number of tiles necessary to cover this room. If the living room area is 168 square feet and the single tile area is one square foot, that says that it takes 168 tiles to cover the floor in this room.

It is important that you impress upon the students' minds that area will be used to discuss the size of surfaces such as decks, carpets, and so on. One usually buys carpet in terms of square yards where the basic square measures one yard on each side. People will talk about the earth's surface in terms of square miles which says the square to which we are referring is one mile on each side.

Circle: As with the rectangle, we are interested in finding the length in traversing the outside of the circle. We call it circumference instead of perimeter. It is found that the circumference C is related to the radius ($r \equiv$ distance from the center of the circle to its boundary, seen in Figure 5.36) by the equation C = 2πr, where π is the ratio of the circumference to the diameter. Also, however, no matter how large or small a circle is, π is a constant approximately equal to $\frac{22}{7}$.

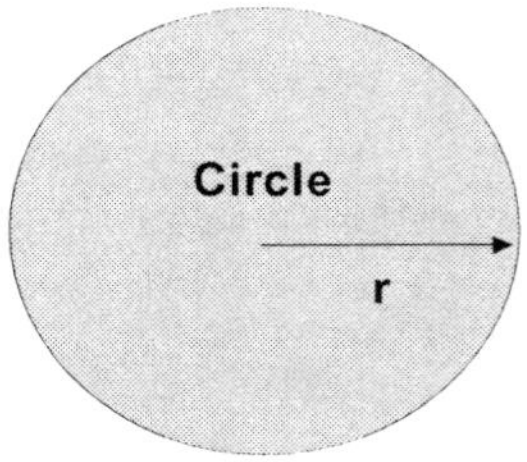

Students can also calculate the area of this circle by first understanding what units (inches, feet, meters, and so on) r will have. The equation for area (A) is A = πr². We get square units here that tell the students that the same square block that they generated in looking at a square also applies in this situation. If they calculate 140 square feet, this says it takes 140 tiles that are one foot square to fill the circle. It should be pointed out that some of the blocks would need

Figure 5.36

cutting because of the curvilinear shape of a circle versus the square shape of tiles.

Triangle: The formula for the perimeter of a triangle is the sum of the measure of its legs. The area of a triangle is one-half of its altitude (a perpendicular line from a vertex to an opposite side of a triangle as is h in Figure 5.37) times its base. The equation for finding the area (A) of a triangle is $A = \frac{1}{2}bh$.

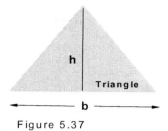

Figure 5.37

Cube: The discussion of the cube (Figure 5.38) offers an opportunity to help students understand that we live in a three dimensional world; that things they see everyday have length, width and height.

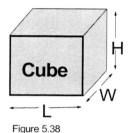

Figure 5.38

A simple experiment that helps students grasp this three dimensional concept is to have a person stand at some point away from a corner in the room and then describe what it takes to reach the tip of his or her nose relative to the walls and floor in the room. Let us call the length along one wall x, and y for the other, and the distance we rise up off the floor z. Go through the movements along x on the floor to where you are aligned with the person, then move along y on the floor until you are just under the person's nose. (It becomes obvious that you must rise some distance off the floor to reach the person's nose.)

At this point the students are ready to accept that every point in the room can now be described in terms of (x, y, z). You can use the walls in the room and the floor to point out that you have a six-sided container when you look at a room. It has a floor, ceiling, and four walls. But how does one know how big the room is?

The size of a room has some meaning when we come up with a unit block such as in Figure 5.38 where all of the sides have a measure of one in whatever units we choose to use. We define the volume of this cube as its length (L) times its width (W) times its height (H). Since all of the sides have a measure of one, then some common units of volume are cubic feet (ft^3), cubic yards (yd^3), cubic centimeters (cc) and cubic meters (m^3). It should be noted that the units are cubic because we add the power on them in multiplication as we do for any algebraic expression.

Rectangular Solid: When returning back to our room model again, chances are it will be a rectangular solid as in Figure 5.39. We are able to get a measure of a rectangular solid's volume by multiplying its length times width times height and then dividing that value by the volume of a cube with the same units. You want to remind the students that the same units must be used on the length, width, and height when making calculations. If they are given a height of six feet and width of five yards and length of six yards, they must convert everything either to feet or to yards before they make their volume calculation.

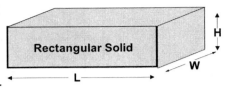

Figure 5.39

Sphere: I find it instructive to use the circle as the basis for a discussion on spheres. I take a coin from my pocket and hold it to show that it is a circle. Then I spin this coin on a desktop and the students see it forms a ball or a sphere. This experiment relates the ideas in the circle to the sphere and now you can say that you calculate the volume (V) of this sphere by the formula $V = \frac{4}{3}\pi r^3$.

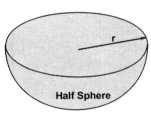

Figure 5.40

Figure 5.40 shows half of a sphere where the students can see the radius r. You must point out that radius r is subtended from the center of the sphere and it touches the inner spherical boundary when it sweeps about.

Figure 5.41 gives us a look at the sphere. When you discuss the volume of this sphere the students ought to be reminded that they must consciously decide the measuring unit system in which they wish to work. Once they calculate their spherical volume, they are now ready to divide it by a unit volume to gain a sense for how large their sphere really is.

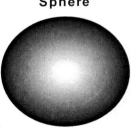

Figure 5.41

Because some students might protest that you cannot put cubic blocks into a sphere evenly, you might need to stress that these blocks could be etched to fit the contour of the sphere. That is, you might make the blocks from a soft material that can easily be shaped into the curvilinear path of the sphere at its boundary.

Circular Cylinder: The circular cylinder (Figure 5.42) is an opportunity to use thinking learned in two arenas. It follows the prescription of determining the area of a circle, that is, its top or bottom, and multiplying that value by the height (h) of the cylinder. The formula for the volume of a circular cylinder is $V = \pi r^2 h$.

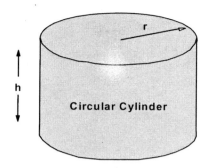

Figure 5.42

You want to remind the students that r and h must have the same units. Also that once they decide upon a set of units, they will have a sense for how big this circular cylinder happens to be.

We have looked at a few key expressions that have everyday utility that students also need to be able to solve many word problems found in their textbook. It is imperative that *you constantly remind students that they need to make sure they are working in a common set of units when attempting to calculate area or volume.*

It is also important here to stress that students can calculate dimensions in cases where they are missing versus attempting to measure them. If the student knows the number of cubic feet in a sphere, he or she can calculate its radius. Students can also use their reasoning ability to calculate the volume between two spheres when they know the design radii.

Once you complete this brief brush with geometry, you want to give the students some spontaneous problems that utilize these principles. If possible, find an experimental way to look at these concepts. Pose situations such as buying tiles to cover a kitchen floor. How many tiles would it take? You might want to bring in a box of tiles and measure out an area, then cover it.

Real life experiments with the mathematics etch indelible marks on students' memories that allow a basis for building more complex material in later discussions.

Graphic Solutions for Functions or Relations

Students may question the need for this learning to manually plot graphs in this age of graphing calculators and graphing utilities on personal computers programs.

Chapter 5

As a teacher, I go straight at the students and state that this material allowed people to trace curves of complex functions with the use of only a few points in the days before computers and graphing calculators. I also believe that *one needs to work with the data to gain valuable insights, so that he or she has a feel for what makes sense.*

This forthrightness with the students tends to ameliorate the need to study graphic solutions without the aid of graphing utilities and graphing calculators in the early portions of this discussion. You want to stress that you want the students to be able to sit in a meeting and listen to a presenta-

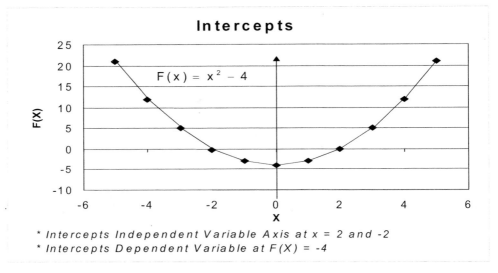

Figure 5.43

tion and draw conclusions based upon expressions made by the presenter in business or scientific presentations.

The goal here is to be able to trace a curve from where it crosses the independent variable axis and the dependent variable axis. We also need to know where it has relative maximums and relative minimums.

Challenge the students to recognize that when the *dependent variable equals zero, that means that the graph is on the axis of the independent variable.* If x is the independent variable and y is the dependent variable, then $y = 0 = f(x)$ gives the values of x where the graph crosses the x axis. This point may not be obvious to some students on the first discussion; therefore, you want to go through the class and make certain that people comprehend it. If you find many people missed the concept, you might consider plotting a simple graph such as $y = f(x) = x^2 - 4$ as shown in Figure 5.43 (Intercepts). Students find in this example that when $y = 0$, then $x = \pm 2$.

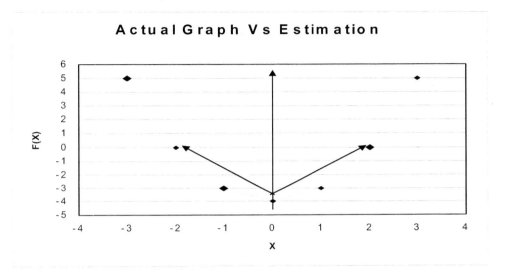

Figure 5.44

Figure 5.43 shows that this graph crosses both the x and y axes. In addition to seeing that $x = \pm 2$ when $y = 0$, the students also see that at $x = 0, y = -4$. Point out that given these interception points, you have an idea of what the graph of this equation looks like. Once you find the interceptions of both the independent and dependent axes, you can make a rough pictorial representation of the curve they will generate.

Figure 5.44 (Actual Graph vs. Estimation) compares our estimated curve shape (denoted by the arrows) to the actual shape when looking at points near the interception points. Although the arrows are straight lines and miss the curvilinear shape of the actual graph, we do get a rough feel for what this graph will actually look like without a great deal of plotting. You should encourage all of the students to go through an exercise similar to the one just outlined because they will have a sense of where you are going as the intercept concept develops in the upcoming lectures. If you feel it necessary, call several students to the blackboard to work out problems you offer off the top of your head.

138   Chapter 5

It is now time to offer the formal definition to cover these intercept situations:

1. "The point (a, 0) is called an x-intercept of the graph of an equation if it is a solution of the equation. To find the x-intercept(s), let y = 0 and then solve the equation for x.

2. "The point (0, b) is called the y-intercept of the graph of an equation if it is a solution point of the equation. To find the y-intercept(s), let x = 0 and then solve the equation for y" (Larson, Roland & Hostetler, p. 180).

Although we limited our discussion to finding the interception of graphs with the independent and dependent axes, it is also important to understand what happens when two graphs cross each other.

A corporation is very interested in finding its breakeven point. This calculation is the interception of the cost graph and the revenue graph.

Figure 5.45 (Interception of Two Curves) offers us a look at two intersecting graphs of two different functions. With close examination of the axis, you can estimate the points of interception. These functions are

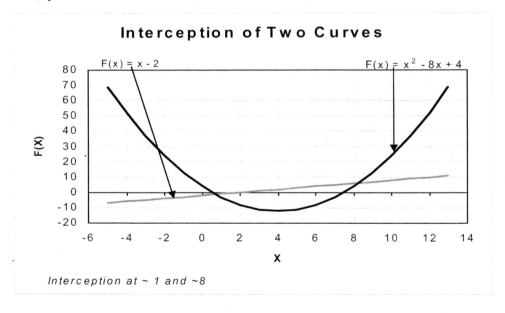

Figure 5.45

First Semester Pre-calculus Under a New Paradigm

$F(x) = x^2 - 8x + 4$ and $F(x) = x - 2$. The two curves intercept at roughly $x = 0$ and $x = 8$. If the students substitute these values into the two equations, they find f(x) equals -1 and -3 respectively in the case where $x = 1$. The students find f(x) equals 6 and 4 respectively when $x = 8$. These significant differences suggest that our graphic solution is not fine enough to pick up the values that offer interceptions of the two curves.

Since difficulty arises in Figure 5.45, it is a valuable exercise to work out the interception points algebraically and then reexamine the graph. Have a student go to the blackboard, set the two equations equal to each other, and solve for the values of x. Here you want to point out that if the two equations intercept then the values for x and f(x) are the same for both equations with these intercepting coordinate values.

The student might offer the following:

$$x - 2 = x^2 - 8x + 4$$
$$0 = x^2 - 8x + 4 - x + 2$$
$$0 = x^2 - 9x + 6$$
$$x = \frac{-(-9) \pm \sqrt{(-9)^2 - 4*1*6}}{2*1}$$
$$x = \frac{9 \pm \sqrt{57}}{2} = \frac{9 \pm 7.55}{2}$$

$x = 0.725$ or $x = 8.275$

You might encourage the student to use the quadratic equation if he or she does not come to this conclusion in a reasonable period. Using the quadratic equation, the student generates the solution for x as $x = 0.725$ or $x = 8.275$. The student needs to substitute these two x values back into the original equations to see if they generate the same f(x) values for both equations. Thus, he or she may offer:

Case I: $x = 0.725$

$0.725 - 2 = -1.275$

$(0.725)^2 - 8(0.725) + 4 = -1.274375$

It is obvious that the f(x) values for both equations are approximately the same.

Case II : x = 8.275
8.275 - 2 = 6.275
$(8.275)^2 - 8*8.275 + 4 = 6.275625$

It is also obvious that the f(x) values for both equations are approximately the same.

The algebraic solution is clearly superior to our graphic solution. However, in cases where one wants to show data in open forums and not divulge corporate secrets, graphic representations without giving exact equations and using gross scales as in Figure 5.45 are excellent vehicles. You want to point this idea out to students to prevent their experiencing unnecessary stress over their inability to find exact solutions with graphs without spending considerable effort to get their scaling refined.

The students should also understand that the above solutions come from the fact that the two equations are set equal to each other and then combined to result in a combination that equals zero. They found the zeros of this combined equation and it yielded the intercept points.

Function vs. Quadratic Equation

The problem with what has been done so far is that we have not derived the quadratic equation although a student used it to solve our graphic interpretation problem. You might broach the quadratic equation discussion by posing the following question. Given a function of the form $y = f(x) = ax^2 + bx + c$, what values of x will tell you that the equation crossed the x-axis? Get the students to discuss. You might offer hints like *what is the value of f(x) when you are touching the x-axis?*

Once the students arrive at the response of f(x) = 0 along the x axis, you are now ready to derive the quadratic equation. You want to state that your real hidden agenda is to solve for those values of x where the f(x) = 0 in order to identify the place(s) where the graph will cross the x-axis. This condition says the students want to solve for x in the expression $0 = ax^2 + bx + c$, which is called a quadratic equation.

It is instructive to walk the students through each step in the derivation and stay away from using expressions, such as "this is intuitive." Statements such as this suggest that you do not want to make your lecture user-friendly and students might conclude that you do not care to explain things. *You want to guard against allowing a series of these subtle "I don't care" messages to cumulate in your students' psyche where they form a bad impression of your teaching style and you find yourself disconcerting to many students as the semester or session progresses.*

Let us talk through the quadratic equation derivation step by step as you might present this material to your students.

1. Our first step is to seek the x values where $f(x) = 0$. This means we set $0 = ax^2 + bx + c$

2. The students need to clear the coefficients from the x terms wherever possible. This suggests that students want to divide both sides by "a" as $x^2 + \frac{b}{a}x + \frac{c}{a} = 0$

3. Students want to further isolate x, by moving the constant term to the other side of the equal sign. This action has all of the x terms on one side $x^2 + \frac{b}{a}x = -\frac{c}{a}$

4. You want to tell the students that what you want to do is to create a perfect square. This perfect square will give us something that we can factor and let us operate on x. The rule for completing the square is to take one-half of the coefficient of x and square it and add the result to both sides of the equal sign as $x^2 + x\frac{b}{a} + \left(\frac{b}{2a}\right)^2 = -\frac{c}{a} + \frac{b^2}{4a^2}$

5. We remove the parentheses $x^2 + x\frac{b}{a} + \frac{b^2}{4a^2} = -\frac{c}{a} + \frac{b^2}{4a^2}$

6. We factor the perfect square. We show $\left(x + \frac{b}{2a}\right)^2 = -\frac{4a}{4a}*\frac{c}{a} + \frac{b^2}{4a^2}$.

There are two points that need addressing here. Many students will struggle with seeing a factor with a fraction and they need convincing that it does square back to the original equation. You want to get a student to go to the blackboard to square the term in the parenthesis using the FOIL method or the long hand multiplication technique. The second issue is in developing a common denominator for the right-hand side of the equal signs where we multiplied the first term by 1. That is $\frac{4a}{4a}$.

7. Put terms on the right side of the equation over a common denominator $\left(x + \frac{b}{2a}\right)^2 = \frac{b^2 - 4ac}{4a^2}$

8. Take the square root of both sides. The students must bear in mind that they get a plus or minus sign in front of the radical because there are two values for x. The resulting equation is $x + \frac{b}{2a} = \frac{\pm\sqrt{b^2 - 4ac}}{2a}$

9. We move the constant term to the right-hand side $x = \frac{-b \pm \sqrt{b^2 - 4ac}}{2a}$, which results in an expression that offers solutions for x.

Complex Numbers

As students start to study the radicand in our solution for the quadratic equation, it becomes apparent that there will be some values where $b^2 - 4ac < 0$. That says that these solutions are undefined based on the present knowledge. However, we know that $\sqrt{-9} = \sqrt{-1}\sqrt{9} = 3\sqrt{-1}$. We can define $\sqrt{-1} = i$ which gives us a system for handling negative numbers under the square root.

The concept of i allows the definition of a new type of number that we shall call an imaginary number that is composed of a + ib where a and b are real numbers and b is not equal to zero. (If b = 0 then we only have a real number a.) If a = 0 and b ≠ 0, we have the imaginary number ib.

We also get the added property for imaginary numbers that if a + ib = c + id then a = c and b = d.

You want to highlight that you can also add and subtract imaginary numbers. However, we must add the real components to real components and imaginary components to imaginary ones. This means that $(a + ib) + (c + id) = (a + c) + (b + d)i$. A similar scenario holds true for subtraction. That is, $(a + ib) - (c + id) = (a - c) + (b - d)i$.

A complex number plus its additive inverse is zero. The students see this concept in the addition of $a + ib + (-a - ib) = a - a + ib - ib = 0 + 0i = 0$ following the above rules on subtraction.

We can also multiply and divide complex numbers. Have a student go to the blackboard and multiply

$$(a + bi)(c + di) =$$
$$a(c + di) + bi(c + di) =$$
$$ac + adi + cbi + bdi^2 =$$
$$ac - bd + (ad + cb)i.$$

Finally, you want to introduce the concept of the conjugate. The complex number $a + bi$ has a conjugate $a - bi$ such that

$$(a + bi)(a - bi) =$$
$$a^2 - abi + abi - b^2i^2 =$$
$$a^2 + b^2.$$

Note that $a^2 + b^2$ is a real number.

General Treatment of Quadratic Equations

Since we now know the quadratic equation formula, students can handle all sorts of quadratic equations where they are looking to find the interception points. Knowing the interception points plus understanding the definition for locating the vertex of the graph of a quadratic function, with $x = -\frac{b}{2a}$ being the coordinate of the vertex, the students now are ready to draw a rough graph from a few points.

Some issues the students must examine are the coefficient of the x^2 term, and whether they will find real and/or imaginary values for x. If we have ax^2 and $a < 0$ then the vertex will occur at the maximum. If we have a > 0 then the vertex will occur at the minimum. We see these situations in Figure 5.46 (Positive & Negative Coefficient Plot) where we are plotting the two functions for f(x) [$f(x) = x^2 + 7$ and $f(x) = -x^2 + 7$] on the

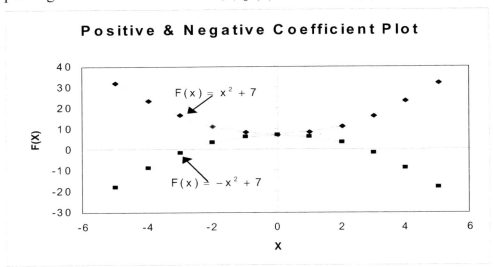

Figure 5.46

same coordinate system. Figure 5.46 shows the positive coefficient for x^2 gave a minimum at the vertex and the negative coefficient gave a maximum at this location.

Figure 5.46 also allows the students to see what happens when they solve for the x and f(x) intercepts and shows the students that they can precisely locate the vertex. In the case of the vertex, we know it occurs at the value $x = -\frac{b}{2a}$. In Figure 5.46 the students see that b = 0, that says the vertex in both graphs occurs at the point x = 0.

In the case of the equation $f(x) = x^2 + 7$, the students see that there are no real solutions for x at f(x) = 0. Its solutions are $x = \pm\sqrt{7}i$. Since there are only imaginary solutions, this tells the students that there are no x intercepts. That point is clear in Figure 5.46. But if you set x = 0, the students find that there is an interception with the f(x) axis at 7.

In the case of the equation $f(x) = -x^2 + 7$, the students see that there are real solutions for x at f(x) = 0. Its solutions are $x = \pm\sqrt{7}$. The students also see that there is an intercept on the f(x) axis at x = 0.

Higher Degree Polynomials

When the degree of the polynomial is greater than 2, you want to now offer students some techniques for finding the interception points with these equations. You might want to point out to the students that if a*b*c*d ....*z = 0 then this implies that some or all can be equal to zero. This permits us to factor our polynomial and set its factors equal to zero to determine the x-axis interception points.

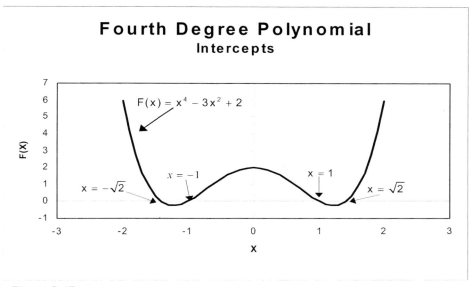

Figure 5.47

Larson, Hotstetler and Edwards offer some examples that hit the key points on higher polynomials (Larson, Roland & Hostetler, p. 205–208). Consider plotting the equation $F(x) = x^4 - 3x^2 + 2$ (Figure 5.47 Fourth Degree Polynomial Intercepts). You would want to start your effort by trying to find the x-axis interception points. Ask yourself, "What are the factors of f(x) that can be set equal to zero?" How to find these factors becomes the pivotal question.

We will introduce two techniques for broaching this problem. If the students think about it, they will conclude that only integer factors of two will be our first choice. You want to bear in mind that factors do not have to be solely integer values. In our above equation we see the factors of 2 are

± 1 and ± 2. This means that F(x) should be divisible by (x - 1), (x + 1), (x - 2), or (x + 2). The students' goal is to find out which of these potential factors are indeed factors of F(x).

What you want to do is to divide F(x) by each potential factor to determine by which ones it is divisible. The students will set each factor equal to zero to determine the interception points.

This gives us an opportunity to discuss long division in algebra. I find it helpful to first go through a numerical long division problem on the blackboard to ensure that everyone is on board for the lesson. Let us examine $24 \overline{)7800}$.

You might place emphasis on the fact that one needs to change the sign when you are subtracting one line from another because many people may struggle seeing this done on algebraic expressions. The details of the division are:

$$\begin{array}{r} 325 \\ 24{\overline{\smash{\big)}\,7800}} \\ \underline{72} \\ 60 \\ \underline{48} \\ 120 \\ \underline{120} \end{array}$$

The students are ready to look at dividing algebraic expressions. The key issue that causes consternation for some students is recalling that $\dfrac{x^n}{x^m} = x^{n-m}$. This definition says that $x \overline{)x^4}$ gives x^3. Once the students feel comfortable with the need to subtract the powers in division, then they can follow the same subtraction idea that they used in the above numerical division problem. However, I encourage my students to *circle the signs that they changed*, so that they can review their efforts quickly.

Now we are ready to return to our original problem of finding the factors:

Case I: (x - 1)

$$\require{enclose}\begin{array}{r}x^3+x^2-2x-2\\x-1\overline{\smash{)}x^4+0*x^3-3x^2+0*x+2}\\\underline{x^4-x^3}\\x^3-3x^2\\\underline{x^3-x^2}\\-2x^2+0\\\underline{-2x^2+2x}\\-2x+2\\\underline{-2x+2}\end{array}$$

Case II: (x + 1)

Since we already have one factor, we do not need to return to the original equation because we know a*b*c = a(b*c). This understanding allows us to hold *a* to the side and examine b*c. We therefore need only divide

$$\begin{array}{r}x^2-2\\x+1\overline{\smash{)}x^3+x^2-2x-2}\end{array}$$

Case III: (x² - 2)

At this point, you only need the students to solve the algebraic expression $0 = x^2 - 2$, which is $x = \pm\sqrt{2}$. You now have all of the factors of the original equation. These factors are $x^4 - 3x^2 + 2 = (x+1)(x-1)(x+\sqrt{2})(x-\sqrt{2})$. These factors tell the students that the graph intercepts the x-axis at the points $x = \pm 1$ and $\pm\sqrt{2}$. The x = 0 term tells us that the graph crosses the f(x) axis at y = 2.

Figure 5.47 shows our x and F(x) interception points. The students can see that the graph matches our calculated values.

Synthetic Division

We used standard division techniques to determine our factors but we can also use synthetic division to achieve our goal. Synthetic division is a technique where one only worries about the coefficients in the polynomial and places zeros where terms are missing.

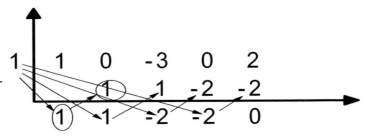

Figure 5.48

We will continue to use our equation $F(x) = x^4 - 3x^2 + 2 = 0$.

An issue to resolve very quickly is to help the students understand that choosing a factor like x - 1 is really saying that x = 1. In synthetic division we want to use the actual values of x we think are the solutions to our problem when deciding upon values to use in our division.

If we are dividing F(x) by x - 1, this means we must divide 1 into the coefficients of the equations. You must put zeroes where terms are missing. Set up your division as in Figure 5.48.

In Figure 5.48 (Synthetic Division), the 1 outside of the arrows is your divisor. The first inside row are the coefficients from your initial equation.

To get the numbers in the second and third rows we will employ multiplication and addition. You bring down the first term in the first row to the third row. You multiply this first term by your divisor, place the result under the second term on the second row, and then add these values and place the results on the third row. In short you multiply on the down diagonals and on the up diagonal place the results under next term on the second row and then add first and second row values.

If the last value on the third row is a zero then the divisor is a factor of your expression. If this number is not a zero, you have a remainder and your divisor is not a factor of the expression under consideration.

Existence Theorems

There is always the lingering question in our minds of whether there are solutions to our expressions. If we have an inkling on the existence of these solutions, we know whether or not to invest our time seeking them. When we also come up with potential factors of our expression, we need some way to narrow our selection, else spend a great deal of time going through each individual one.

The key issue is if an expression $x - k$ is a factor of our expression then $x - k$ will divide into our expression without a remainder. This is the same idea we employ when we say 36 is divisible by 12 because it has no remainder. Thus, $f(k) = 0$ if our expression is a factor of our polynomial. This tells us that if $f(x) = x^2 - 2x - 35$, we find that its factors are $(x + 5)$, $(x - 7)$; hence, its k values will be $k = -5$ and 7. Our theorem tells us that $f(-5) = (-5)^2 - 2(-5) - 35 = 0$.

We have looked at finding integer factors of polynomials, but not all divisors are whole numbers. Some are rational numbers of the form $\frac{p}{q}$ where $q \neq 0$. If we are looking for the interception points of our graph with the x axis, the zero test tells us,

"If the polynomial $f(x) = a_n x^n + a_{n-1} x^{n-1} + \ldots + a_2 x^2 + a_1 x + a_0$ has integer coefficients, every rational zero of f has the form Rational zero $= \frac{p}{q}$ where p and q have no common factors other than 1, p is a factor of the constant term a_0, and q is a factor of the leading coefficient a_n." (Larson, Roland & Hostetler, p. 273).

This statement tells us that in addition to looking at numbers that are divisible into a_0, we can now add to those numbers another set that is made up of Possible rational zeros $= \frac{\text{Factors of } a_0}{\text{Factors of } a_n}$. For example, if the students are seeking the zeros in an equation, such as $2x^3 - x^2 - 37x - 35$, then they will have the following values:

$$\frac{\pm 1, \pm 5, \pm 7, \pm 35}{\pm 1, \pm 2} = \pm 1, \pm 5, \pm 7, \pm 35 \pm \frac{1}{2}, \pm \frac{5}{2}, \pm \frac{7}{2}, \pm \frac{35}{2}$$

as starting places.

Barnett and Ziegler point out a good method of handling this many potential divisors in their Synthetic Division Tables during their discussion of polynomials (Barnett & Ziegler, p. 293-339).

I will state some other existence theorems from Barnett and Ziegler because we already have some techniques for finding the factors and we are now looking for direction on our search. We shall look at some ideas to limit the amount of effort we need to employ to obtain our zeros.

Fundamental Theorem of Algebra: "Every polynomial P(x) of degree n > 0 has at least one zero." This tells our students that there will be at least one factor for every polynomial. Therefore, anyone saying they just cannot find the solution and doubting its existence should be made to understand that it will require a bit more work because it does exist.

n Zeros Theorem: "Every polynomial P(x) of degree n > 0 can be expressed as the product of n linear factors. Hence, P(x) has exactly n zeros—not necessarily distinct." The key issue here is that the students can look at the degree of the polynomial and determine the number of linear factors it has. These linear factors may not be distinct and they may not be real. For example, the polynomial $P(x) = (x-3)^3(x+5)(x-4i)(x+4i)$ has repeat and imaginary terms. The number of times a linear factor occurs is called a *zero of multiplicity m*.

Imaginary Zeros Theorem: "Imaginary zeros of polynomials with real coefficients, if they exist, occur in conjugate pairs." The issue here for the students to grasp is the idea that imaginary solutions always occur in pairs and one is the conjugate (if a + bi is a complex number then a - bi is its conjugate) of the other. You may want to go over what conjugate means to be sure that students are following your discussion. It may be helpful if you use the quadratic equation and have a student go to the blackboard and solve the expression $x^2 + 3x + 9 = 0$ for x. The student should find that $x = \frac{-3 \pm 3i\sqrt{3}}{2}$. If the students examine these solutions closely, they see they are conjugates of each other.

Real Zeros and Odd-Degree Polynomials: "A polynomial of odd degree with real coefficients always has at least one real zero." If the students see an odd degree (degree being the highest power in a polynomial of a single variable), then this says that when you outline the possible real factors, you know that at least one of them will work. Consider the expression $x^3 - 9x^2 + 9x - 1 = 0$. Our theorem tells us that there is at least one real zero. We can use synthetic division and rational zero theorem to quickly find the real solution to our expression, which is $x = 1$.

Descartes' Rule of Signs: "The number of positive zeros of P(x) is never greater than the number of variations in sign in P(x) and, if less, then always by an even number. The number of negative zeros of P(x) is never greater that the number of variations in sign in P(-x) and, if less, then always by an even number." Here we have two considerations to examine. Consider our above equation $x^3 - 9x^2 + 9x - 1 = 0$. It can have no more than three positive factors and its factors are $x = 1$ and $4 \pm \sqrt{15}$ where the second two factors come from using the quadratic formula to solve the quadratic equation remaining ($x^2 - 8x + 1 = 0$) after the first term is found by synthetic division. You can see that even functions will not have sign changes when you are considering the impact of replacing x by -x. You will want to make up a few problems to allow the students to see all of the possibilities.

Upper and Lower Bounds of Real Zeros: "Given an nth degree polynomial P(x) with real coefficients, $n > 0$, $a_n > 0$, and P(x) divided by x - r using synthetic division: If $r > 0$ and all numbers in the quotient row of the synthetic division, including the remainder, are nonnegative, then r is an upper bound of the real zeros of P(x). If $r < 0$ and all numbers in the quotient row of the synthetic division, including the remainder, alternate in sign, then r is a lower bound of the real zeros of P(x)." Send a student to the blackboard and give him or her at least a third degree equation that you know will have some real factors to determine these potential factors, then go through his or her synthetic division. This time you want the students to pay attention to the sign pattern on each quotient row they generate. If all of the numbers in this quotient row have positive signs, then that tells them their divisor is an upper bound if it is a positive number. Have the student to continue until he or she runs into the situation where their divisor is negative and they find alternating signs in the quotient row that says they have a lower bound. Once the students know the boundaries, they can discard the potential values that fall outside this region.

Location Theorem: "If P(x) is a polynomial with real coefficients, and if P(a) and P(b) are of opposite signs, there is at least one real zero between a and b." Here you want to point out to the students that if they think about the real number line, there is no way to get from positive numbers to negative numbers without going through zero. The location theorem is saying a similar concept. In order for our expression to generate both positive and negative values in some range, there must be an interception of our graph with the x-axis.

Solving Equations Involving Absolute Value

Before you start on absolute value, it is helpful to remind students of its definition: |a| equals a when a > 0 and - a when a < 0. This definition implies that there are two conditions that must be examined when solving an absolute value problem.

Let us look at the problem $|4x^2 - 5x| = 30x + 50$ where we must examine two cases.

They are $4x^2 - 5x = 30x + 50$ and $-(4x^2 - 5x) = 30x + 50$. We are following the definition and accounting for the possibilities that our argument a > 0 and a < 0.

You want a student to go to the blackboard and work out both of these cases. We shall go through them here.

Case I: $4x^2 - 5x > 0$

$$4x^2 - 5x = 30x + 50$$

We want to bring everything to one side of the equal sign.

$$4x^2 - 5x - 30x - 50 = 0$$

Combine like terms and generate a quadratic equation set equal to zero.

$$4x^2 - 35x - 50 = 0$$

First Semester Pre-calculus Under a New Paradigm

Use the quadratic equation to solve for the x values.

$$x = \frac{35 \pm \sqrt{(-35)^2 - 4(4)(-50)}}{2*4}$$

$$x = \frac{35 \pm 45}{8}$$

$$x = 10 \text{ and } -\frac{5}{4}$$

Case II : $4x^2 - 5x < 0$

$$-(4x^2 - 5x) = 30x + 50$$

We bring everything to one side of the equal sign.

$$-4x^2 + 5x - 30x - 50 = 0$$

Combine like terms; multiply both side by (-1) and form the quadratic equation.

$$4x^2 + 25x + 50 = 0$$

Use the quadratic equation to solve for the x values.

$$x = \frac{-25 \pm \sqrt{25^2 - 4*4*50}}{2*4} = \frac{-25 \pm \sqrt{625 - 800}}{8} = \frac{-25 \pm 15i}{8}$$

There are no real solutions for the negative condition in this absolute value equation, so it does not intercept the x-axis. However, the positive condition for the absolute value offers a real solution.

Solving Equations Involving Fractions

The key issues here are that students understand the need for a common denominator and how to multiply all terms by it to remove denominators on the various fractions terms. Although we shall solve the fraction equation $\frac{3}{x} - \frac{5}{x+2} = -\frac{7}{60}$ this is an excellent problem to have a student go to the blackboard to work out. Your students have advanced to where this sort of calculation should only take minimal student effort.

We shall establish the common denominator as 60x(x+2).

We multiply each term on both sides of the equal sign by the common denominator. We are using the definition if a = b then ac = bc.

We distribute the common denominator.

$$60x(x+2)\left[\frac{3}{x} - \frac{5}{x+2}\right] = -\frac{7}{60} * 60x(x+2)$$

$$180(x+2) - 300x = -7x(x+2)$$

Do all multiplications, rearrange terms, and multiply both sides by (-1).

$$-120x + 360 = -7x^2 - 14x$$
$$0 = -7x^2 - 14x + 120x - 360$$
$$0 = -7x^2 + 106x - 360$$
$$0 = 7x^2 - 106x + 360$$

We now use the quadratic equation to find our solutions.

$$x = \frac{106 \pm \sqrt{106^2 - 4*7*360}}{2*7} = \frac{106 \pm 34}{14}$$

x = 10 and ~ 5.1428

Radical Equations

You want to remind the students of the following:

First Semester Pre-calculus Under a New Paradigm

$$\sqrt{a} * \sqrt{a} = a^{\frac{1}{2}} * a^{\frac{1}{2}} = a^{\frac{1}{2}+\frac{1}{2}} = a^1 = a.$$

You may wonder why I went through each step. I look for every opportunity to reinforce key materials that are necessary to handle advanced subjects.

You need to get a student to work through this one or you may find that you start losing people's attention because your lectures get a bit long-winded. Our problem is to solve $\sqrt{3x+6} - \sqrt{x-1} = 3$.

If the student at the blackboard hesitates, encourage him or her to place one radical on one side where it can be squared easily without generating new radical terms.

$$\sqrt{3x+6} = 3 + \sqrt{x-1}$$

The student should square both sides. You want to point out to the class that the goal is to do whatever it takes to remove the radicals to be able to solve for x.

$$3x + 6 = 9 + 6\sqrt{x-1} + (x-1)$$

The student wants to keep the radical on one side of the equal sign, combine the other terms, and move them to the other side.

$$3x + 6 = 9 - 1 + x + 6\sqrt{x-1}$$
$$3x + 6 - 8 - x = 6\sqrt{x-1}$$
$$2x - 2 = 6\sqrt{x-1}$$
$$x - 1 = 3\sqrt{x-1}$$

Now the student can square both sides of the equal sign and generate an equation that we can use the quadratic equation to solve.

$$x^2 - 2x + 1 = 9(x-1)$$
$$x^2 - 11x + 10 = 0$$

Since we have a quadratic equation, the student should demonstrate his or her ability to use the quadratic formula to solve for the x values. This person should find that

$$x = \frac{11 \pm \sqrt{11^2 - 4*1*10}}{2*1}$$

$$x = \frac{11 \pm \sqrt{121 - 40}}{2} = \frac{11 \pm 9}{2}$$

x = 10 or 1

Although the student may feel good that he or she solved the problem, you want to challenge the students to return to the original equation to input their answers to see that they are truly the solutions they seek. They should see the following:

When x = 1

$$\sqrt{3*1+6} - \sqrt{1-1} = \sqrt{9} - \sqrt{0} = 3$$

When x = 10

$$\sqrt{3*10+6} - \sqrt{10-1} = \sqrt{36} - \sqrt{9} = 3$$

The above calculation makes clear that the students have indeed solved the equation and this effort should become a routine part of their effort in seeking solutions to various equations.

Note the general rule on conversion from rational exponents to radicals is: $x^{\frac{p}{q}} = \sqrt[q]{x^p} = \left(\sqrt[q]{x}\right)^p$

Properties of Inequalities

Transitive Property: If a < b and b < c, then a < c. An example of this using money is the one dollar bill is less than the five dollar bill and the five dollar bill is less than the ten dollar bill. This says that the one dollar bill is less than the ten dollar bill, which is common sense on the part of the students.

Addition of Inequalities: If a < b and c < d, then a + c < b + d. An example using money is if the student has five dollars in one pocket and ten dollars in another pocket and he or she decides to add seven dollars to the first pocket and twenty to the second, the money in the first pocket (twelve dollars) will be less than the thirty dollars in the second pocket.

Addition of a Constant: If a < b then a + c < b + c. An example using money is, if the student has fifty dollars in one pocket and sixty dollars in a second pocket and she or he adds twenty-five dollars to both pockets, the amount in the first pocket will still be smaller than that in the second pocket.

Multiplying by a Constant: a < b implies ac < bc when c > 0, but ac > bc when c < 0. It is important to go through why you flip the inequality sign because many students may not initially see this need. A good way to make this point is to set up an inequality like 3 < 9. Then multiply both sides of the inequality side by -2: $3*(-2) < 9*(-2)$. When the students look at your attempting to say that -6 is less than -18, this assertion will run counterpoised to what they learned about the position of numbers on the real number line. Students know that -18 is less than -6, therefore, they will see the need to flip the inequality when multiplying by a negative value, meaning that -6 > -18.

The standard technique is to evolve our discussion into one where we start to examine inequality equations. The key problem with that method for business and social science students is you have offered no real reason for studying this material other than mathematical curiosity. Therefore, you ought not be surprised if your students show little passion for what you are attempting to offer.

I find it helpful to pose a real world problem to the students that employ the techniques of inequalities and have the students write out on the blackboard their various equations to describe the situation we have under study. I say something such as, "You are working at a credit card company and you all want to issue a new platinum card to people between the ages of thirty and forty-five with a picture of a woman wearing a mink coat standing in front of an expensive luxury automobile. What criterion should we use to find people in the company's current customer database to issue the new platinum credit card?"

Call a student to the blackboard to write down the requirements from the class. This student should first write down the problem, so there is no debate over what they are asked to do. You want to have this student write a mathematical expression for each requirement as the class offers it. You also want the student to draw a diagram of the mathematical expression, so that the class has a pictorial representation of each condition.

* People must be at least thirty years of age and not over forty-five. This indicates an equation of the form $30 \leq x < 45$ or the set $[30, 45)$. Figure 5.49 (Double Inequality) offers us a pictorial representation of our double inequality.

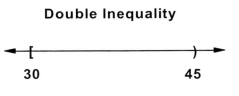

Figure 5.49

* The customer must make over 50,000 dollars per year. The salary is represented by the equation $y > 50000$ or the set $(50000, \infty)$. Figure 5.50 (Salary) shows a graph of this element in the customer requirements the students are developing.

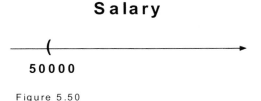

Figure 5.50

* Have a credit rating of at least four on a scale of one to five. The equation is $r \geq 4$. Figure 5.51 (Credit Rating) is the pictorial representation of the credit rating. However, you want to ask the students if they feel they have sufficient information to make a decision on a person getting a platinum card at this point. It is important to see if they are assessing their situation or merely pulling things out of the clouds. You should not settle for an answer where the students merely say, "I don't know." One of your hidden agendas in this exercise is to help students gain confidence in their own abilities to make

decisions based on facts and not succumb to merely emotional reactions to situations.

    * It is important here that you get the student at the blackboard to make some sort of rough pictorial representation of the credit card with the woman wearing the fur coat standing in front of the luxury car. When students make pictorial representations, they are forced to think about what they are attempting to achieve. Once the picture is completed, have two students write a mathematical expression to represent our problem. If the students do not immediately attempt to write an expression representing the lady wearing the mink-fur coat, you need to remind them that you are looking for a yes or no expression. Do not allow the students to skip this exercise because it will prevent a real world debacle. Imagine the reactions from animal rights activists or animal lovers who may resent seeing a woman wear a fur coat. One possible solution to this problem is to let $h \neq a$ where h is hobby in the customer profile and a is credit card company code for animal lover.

    There are no doubt more elements the students could put down, but now you want to ask them if they feel they have sufficient information to make a decision. They will most likely feel some confidence that they can make a prudent business decision on this credit card issuance. Students should now see that this exercise made the inequality concept take on relevance and they will be receptive to this new concept you are going to present.

    The key issue in all of the responses we found above is that there is no discrete answer to our inequality; it has a host of values that make the statement true. In each condition above, we constricted the population based on the various filters we used. The students should be made to see that first we took the universal set, which was all of the credit card company's customers, and we took a subset of people greater than or equal to 30 but less than 45 (Call this subset A). We found the subset of A that makes 50,000 dollars or more (Call this subset B). From a subset of B we found all of the people with a credit rating of 4 or better (Call this subset C). The final step is subtract out of subset C all the customers who are animal rights enthusiasts, which gives us our target customer set.

    We will now walk through a couple of inequality problems that touch on some key points on this subject. You can talk the students through each step.

Problem 1

$$2 - \frac{4x}{5} > 2x + 9$$

$$5 * [2 - \frac{4x}{5} > 2x + 9]$$

$$10 - 45 > 10x + 4x$$

$$10 - 4x > 10x + 45$$

$$-\frac{35}{14} > x$$

$$-\frac{5}{2} > x$$

Problem 2

$$|x - 9| < 10$$
$$-10 < x - 9 < 10$$
$$-1 < x < 19$$

Problem 3

$$x^2 - 2x > 35$$
$$x^2 - 2x - 35 > 0$$
$$(x + 5)(x - 7) > 0$$

The critical points are x = -5 and x = 7. We want to test values around these points to see what makes sense to use. There are three intervals to evaluate: $(-\infty, -5), (-5, 7)$, and $(7, \infty)$. If we select the x value of -6 from the first set, we find that it yields 13 > 0. If we select -4 for the x value in the second interval, we find that it yields -11 is not greater than 0. Finally, if we select 8 for the x value in the third region, we see it yields 13 > 0. These exercises suggest that the values of x that satisfy the inequality are $(-\infty, -5) \cup (7, \infty)$.

Problem 4

$$\frac{5x-4}{2x+5} < 4$$

$$\frac{5x-4}{2x+5} - 4 < 0$$

$$\frac{5x-4-4(2x+5)}{2x+5} < 0$$

$$\frac{-3x-24}{2x+5} < 0$$

The expression critical points in Problem 4 are x = -5/2 and x = -8. The three intervals of concern are $(-\infty, -8)$, $\left(-8, -\frac{5}{2}\right)$, and $\left(-\frac{5}{2}, \infty\right)$. The first interval meets the inequality choosing the point x = -9, the second interval fails selecting the point x = -7, and the third point meets the expectation selecting the point x = 0.

Summary

When you think about being able to radically improve business and social sciences students' abilities to participate in research and understand the forefront findings in their professions, it becomes obvious that these students need to understand pre-calculus I. In business degree programs, students will have to study subjects such as finance, finite mathematics, statistics, and business calculus to gain the knowledge necessary for sound business decisions. It is here we see that pre-calculus I is the gateway course. Without pre-calculus, business students are unable to fully share in the knowledge that helps them become world-class operatives in many modern business arenas.

However, pre-calculus I presupposes an understanding of arithmetic and it is here you must assess the students' preparation and develop your course accordingly. Do not be surprised if students lack knowledge of the

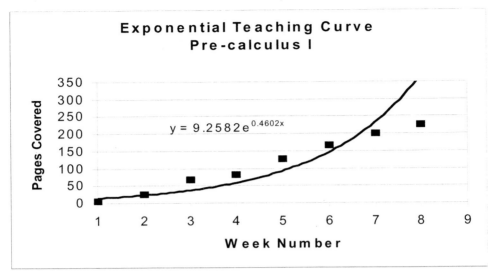

Figure 5.52

times tables and are unable to do fractions and decimals. This means that you must fill in background holes before you start your pre-calculus I journey if you hope to carry the overwhelming majority of the class with you throughout the course.

Addressing problems of students with poor background preparation, I find the model of teaching according to an exponential model of the form $y = cA^t$ can be very helpful in making these students become successful. You can get a handle on the constants by assessing the capability of the class at the first meeting and making up a new or modifying a standard course syllabus. Then take this syllabus and plot the pages of material covered weekly or biweekly versus the number of weeks in a semester or session using a graphing routine. Figure 5.52 (Exponential Teaching Model) shows the plot of the pre-calculus I syllabus that this writer used to teach courses in Delaware State University's eight-week session course. The equation for Figure 5.52 is approximately $y = 9.26e^{0.46x}$.

You may want to modify this curve as the semester or session progresses as you understand the class's capability better. That is, *the syllabus curve is merely your best guess based on teaching experience and the departmental requirements for a satisfactory course.*

You recognize that pre-calculus I is the underpinning of advanced business and scientific work that requires an understanding of higher level mathematics such as statistics, trigonometry, finite mathematics, linear algebra, calculus, differential equations, and so on. We, therefore, want to look further at the concepts we employed to teach this course. In the past, this mathematical information was taught to help prepare people to pursue careers in hard science, engineering, and mathematics. This hard science mind-set lends itself to the standard teaching recipe of giving a mathematical definition, going through proofs of theorems, and working a host of problems that were either scientific oriented or excellent brainteasers.

In this hard science and mathematics mind-set, a lack of knowledge of pre-calculus I offers colleges and universities an excellent vehicle to screen out potential candidates for admission to their institution. This thinking suggests that universities only admit students who are mathematically mature and ignore the rest of the population who have the mathematical potential but lack the background. Some people will argue that filling in inadequate backgrounds is not the role of the university.

This is a parochial view of the role of the university in helping students with inadequate preparation. The key problem raised in this sterile view comes from the fact that many institutions today have a bimodal distribution of ages in their student body. If we look at 1983, the U.S. National Center for Education Statistics reports that approximately 41 percent of students enrolled in college were twenty-five years old or older (College Enrollment, 1998, p. 188). They also reported in a report entitled, "Special Analysis in 2002 Nontraditional Undergraduates" on the significant numbers of nontraditional students attending college. " There are proportionately more older students on campus as well: 39 percent of all postsecondary students were twenty-five years or older in 1999, compared with 28 percent in 1970 (U.S. Department of Education 2002b)." (Special Analysis 2002 Nontraditional Undergraduates, 2002).

Thus, as a teacher of both young and maturing adults, you should not be surprised to see students in your classroom with two different agendas. Students over 25 may be making a concerted effort to understand the information whilst secondarily seeking a grade, and those under twenty-five may have a reverse objective. This dichotomy means you must devise a plan to focus your students' educational pursuit on understanding the subject material as their primary objective and allow them to see that good grades will be the natural outcome of their effort.

It is imperative that your students understand from day one that your focus will be on understanding and you must maintain this mind-set throughout the course. I establish this understanding mind-set during the first lecture where the times tables are discussed from the point of view of understanding what they really say. This discussion begins by having a couple of students go to the blackboard and write the times tables through the twelve times tables. The initial focus is on helping students to transition from arithmetic to mathematics where students can see such principles as $a * 0 = 0$, $ab = ba$, $1*a = a*1 = a$, and $5*6$ means you have, perhaps, five piles of money with six dollars in each pile. You can then emphasize that your abstract equations hold true for all the examples they can imagine versus each time only solving a specific problem.

At this point, I find it helpful to give the students a proverb. "Algebra is nothing more than high-powered arithmetic where you can solve every problem you can think up once you manipulate the letters then just replace these letters with numbers." This proverb falls naturally out of the lecture on the times tables. It is also important to point out that I give people two weeks to know the times tables if they do not presently have this knowledge, because the students' knowing these tables is a necessary condition for people to learn fractions.

You will note that I used money in my example above. I learned from chatting with chaps that stand on the street corners in Wilmington, Delaware's, 'hood to couch my examples in money. You will come to see that people will understand your comments if you heed the advice of the sidewalk philosophers because many people will see the immediate utility in your lessons when they are couched in terms of money. Think about it, if someone were hungry, why would you want to discuss cutting pies with him or her?

I carry this understanding mind-set a bit further when I have the students gather around my desk and I drop some coins onto the desk and ask one student to count the coins. Many people will give you the value of the coins instead of counting them. Here you have an opportunity to challenge the students to listen to what was asked and not presume what the question would be.

Next, you want to have them give you the value of the coins. Once they offer the coins' value, challenge one or two students to explain exactly how they came to their conclusion. This exercise gives you an opportunity to show them that they already know what a common denominator is because they cannot count the coins until they change all of them to pennies.

You now have a good entrance to discuss fractions. Many students do not understand what a fraction tells them. If you return to our money idea, you might ask someone to tell you how to split 140 dollars equally among seven fellows. You make the point that if the calculation is wrong he or she might expect some adverse reactions. It is a valuable exercise to get them to draw a picture of this problem. Once the students tell you twenty dollars for each person, then you should proceed to tell them that the number of people told you how many piles of money you had to have. If you divide the number of piles into how many dollars you have, then they know how many dollars are in each pile.

Hence, a fraction is made up of a numerator (top number) and a denominator (bottom number). In the fraction $\frac{3}{5}$, the denominator 5 tells you how many piles of money you want to make and the numerator 3 tells the number of these piles you have on hand. If you put several fractions on the blackboard and get every student to tell you what it says, you will accelerate their educational metamorphosis towards adopting understanding as the norm.

As you proceed through the above discussion, you want to assess what percent of your students got through school on rote memorization instead of understanding the material. Many older students came through an educational style where memorization was perceived as students learning the material, since they are good at regurgitating what they hear in class. Yet, students often confess that they had no idea what the teacher really said, but they made it out of the course.

You may find that the use of the blackboard is initially disturbing to students who memorized their way through high school or with people who are hell-bent on sliding through just barely passing lower-level courses. Pre-calculus I is a low enough level course that you ought not expect great resistance when you initiate a student cultivation effort because it is not a major embarrassment to not understand the fundamental concepts from the course prerequisites. However, you can expect resistance from some students in business calculus if they do not fully understand the lower-level courses. Thus, you can save yourself a good deal of grief if you promote understanding in pre-calculus I.

It is also important that you not merely look at your students as youths. You must recognize that significant portions of these people are self-sufficient adults complete with families. That means they are professionals whose skills you are helping to upgrade. Older students must not only learn the classroom material but must also juggle their family responsibilities.

Nevertheless, I find it promotes learning and encourages students to manage their lives well if you do not accept excuses for nonperformance.

Once you establish this no-excuse mentality in pre-calculus I, it carries over into the advanced courses.

In pre-calculus, it is very important to not allow students to put themselves down if they do not understand a given item. I do not allow students to sit down from the blackboard until they grasp the concept under discussion. This makes the blackboard trip a positive experience and it shows the student that success is cumulative.

You will find that many students must be taught success because somehow they formed an enchantment with failure. Low expectations must be driven out of some students' mind-set over the semester, and you accomplish that by making success become the norm. In key areas where I know students will have difficulty grasping concepts, I make each individual student go to the blackboard to demonstrate an understanding of that principle. This technique of seeing that each student understands key concepts will allow you to keep down student attrition resulting from disenchantment due to a lack of understanding.

Although some institutions do not require student attendance in class, I tell my students, *I cannot help you if you are not here. Thus, I give an oral grade that is twenty-five percent of your semester grade and it is earned by your work at the blackboard.* I go on to say that I have not found anyone capable of getting their blackboard assignments done while not present in class. One term I did not enforce this class participation concept and I found that students' performance suffered. These students used excuses in place of performance. Their failure to mature during the semester impeded the progress of the class. Hence, it is imperative that students attend class in gateway courses such as pre-calculus I.

You must teach reading while you are teaching mathematics. If you take each definition and theorem, have a student write them on the blackboard and then talk the students through in layman's terms. You will increase people's vocabulary and teach them how to read a mathematics book. As you move through advanced material, you want to call students' attention back to the basics such as $a = b/c$ if $c \neq 0$ because $c = 0$ is undefined. Of course, students need a picture of what happens when $c = 0$, so you can have them divide a number by zero on their calculator and notice that they see an E for error.

In the early portion of the pre-calculus course, it is very important that you do not allow students to fall victim to excessive use of their calculator. Students need to master addition, multiplication, subtraction, division, fractions, and decimals before they spend a lot of time using a calculator. Otherwise, your students will learn the calculator and not gain an appreciation for the principles of pre-calculus, a scenario that leads to a disaster when they attempt to learn the higher level mathematics courses. This decision will not be popular, but it is here you must make decisions on what is best for the long-term development of the class and not what is expedient now. That is to say, *"Your job is to prepare the students for a lifetime of learning because many professionals will change jobs many times before they retire."*

When students are working at the blackboard, it is especially important in pre-calculus I to see that they learn to work at the highest standards possible. If the student at the blackboard writes something incorrectly, make him or her erase it and put in the correct response. I find it much easier to help students establish good working habits in pre-calculus than have to force them to unlearn bad habits in higher level courses. Once students know your expectations, they usually conform without much fanfare.

Although we have spoken on how to present things to students, it is imperative that your mind set be right for teaching. *You must believe in your heart that all of your students can learn and always teach with that feeling governing your mind-set.* Students know when you care and their responses to your desires will be based on their perception of your commitment to their long-term success. You should not be surprised if students, who have been dormant in other classes, ignite with zeal to succeed when they recognize that you care and will take the necessary time to encourage their learning. Since pre-calculus I is the gateway course, you want to make every effort to kindle the students' passion for understanding here.

Some classroom decorum items I find focus students' attention on learning are: no eating in the classroom, no sleeping in class, no sidebar conversations, no working on other course assignments, students take turns cleaning the blackboard before class starts, and fellows must take off their hats and head rags. These rules never vary throughout the semester and I take immediate action if there are infractions of these rules. For example, people caught in sidebar conversations must share their discussion with the remainder of the class. On the other hand, people caught sleeping must leave the classroom for the day.

A detailed discussion on how to teach the various aspects of pre-calculus I is given in the earlier portions of this chapter; therefore, I will not rehash too much of this material here. Two areas that were not discussed are Partial Fractions and Logarithms. These subjects would fit well into an old paradigm pre-calculus course, but logarithms are discussed in higher level courses such as finite mathematics and there is no pressing need to cover partial fractions in a course where the majority of the student population will be business and social science majors.

When you take a holistic look at what I am suggesting above, it can be summed up in a comment from the *Atlantic Monthly*. It says, "The most important thing in education is what the teacher does with students in the classroom. To direct that requires control of the curriculum. Structural changes, supposedly the essence of the education reform, can have amazingly little effect if they do not alter what teachers actually teach" (Rigors and Routine, 1998). You must take control of what you are teaching and share that with your students through careful thought given in making up a syllabus. Where you must use generic syllabi, consider them as okay for knowing the boundaries of the course material required by your department, but you as the teacher must use your judgment as to how the course progresses to modify the syllabus as necessary.

Since pre-calculus I is the gateway course for the higher level courses, you want to make every effort to instill good habits in students at this level. You must also keep in mind that you are not merely teaching young adults as many people may be your age-peers. Older students have life experiences that can impact on their learning capability. Thus, Kim Hermanson offers us plenty of food for thought when it comes to helping adults learn.

Hermanson writes, "Adults learn by constructing meaning from their experiences. Situations which aren't viewed as meaningful are typically rejected as a source of learning. An important aspect of these meaningful learning experiences is that they not only involve one's intellectual faculties, but one's emotional capabilities as well. Thus, incorporating elements of social and developmental learning will undoubtedly enhance the impact of standard learning activities" (Hermanson, 1998).

If our goal is to teach a cross section of people with varying backgrounds, we must heed Hermanson's wisdom in our pre-calculus I classes and not merely say it is not my job to have to worry about the students' emotional health as well. Consider that pre-calculus I is a gateway course, so we, its teachers, carry the added responsibility to guide as many people as have the talent through this course for the long term strength of our nation.

Chapter 6

Testing

We need to get some assessment of the mind-set of the teachers whom the students had prior to arriving in our classrooms. What we want to do is to set up some personal indicators of how effective our teaching is prior to final examinations so that we can make adjustments to our modus operandi as needed.

Patricia A. Wasley sums up the teacher's mind-set in her article entitled, "Responsible Accountability and Teacher Learning." She writes, "Although most teachers worked hard to ensure that their students were successful, many grew to believe that poor children and children of color do not achieve as well as their more privileged white counterparts because they saw these differences play out in their own classrooms. It is also true that teachers teach the way they were taught themselves... As students, many teachers were taught by teachers who did not believe that all children could learn. The bell curve, then, became a kind of self-fulfilling prophecy influencing teachers around the country for decades" (Wasley, p. 137).

Wasley's comments offer an appreciation between the operational definitions of racially desegregated public schools where all children can legally attend classes, and racially integrated schools where teachers feel a sense of responsibility to develop the minds of all students regardless of their race, creed, color, or national origin. Hence, the day of intraschool de facto segregation is now relegated to a relic of the nation's evolution from a legally racially segregated society to the mainstream receptivity of today's racial diversity as the national norm. Therefore, one's teaching job is complicated by the fact that many students know when teachers have written them off and they act against this teacher disdain by becoming troublemakers.

Wasley offers hope for the future. "New teachers are now taught that every child can learn, provided that the teachers take advantage of the child's prior knowledge and adapt instruction to his or her cultural background and/or pedagogical dispositions. . . The key to being successful with a diverse group of children is that teachers must have the variety of approaches, the depth of subject-matter knowledge, and the diagnostic skills to determine and facilitate what each needs."

Wasley has made the case that a paradigm shift has occurred in the development of tomorrow's public school teacher's mind-set. Thus, as a college professor who is teaching students needing significant background enhancement, you want some simple techniques to manage your classroom to rekindle students' pursuit of excellence and discard their apparent enchantment with failure. I find that students who need background enhancement must have the right balance of classroom learning activities and teacher lecturing. You ought to expect to lose the interest of most of your class if you spend your entire mathematics period lecturing.

Since students usually rise to the level of expectation placed upon them, your job is to set a mainstream performance expectation in the minds of background deficient students. You want to establish the classroom modus operandi the first class and not deviate from it during the semester. These classroom rules may include no sleeping, no drinking and eating, fellows taking off hats, and no harassing fellow students or the teacher. Students are expected to have all of their class assignments completed and you will not accept excuses for nonperformance. If students decide to violate your rules, you must take aggressive action, such as sleepers and students acting in haughty mannerisms must be asked to leave class for the day.

On the other hand, when there is adherence to the classroom decorum by the majority of your class, it offers a gauge on your teaching effectiveness. A couple of students holding full time employment might get tired in class, so you may have to highlight the problem to them the first time and ask them to leave class on the second occasion. But if you find a significant number of students struggling to maintain interest in your lectures, you might assume they are reacting to too much time being spent on lectures or your lectures may need to be more dynamic.

Student Testing

The important issue in student testing is you are going to give the student an opportunity to show her or his understanding of the classroom material in solving real world problems. Skill tests may not be what the students will need in a real world environment where many of their assignments may come via letter, memo, e-mail, presentation, or other written materials. Your students need to know how to do quantitative tests that require them to think issues through and not merely qualitative exercises that show they can solve a complex equation. Hence, you might expect many background deficient students to have a bad day sometime during the semester and you cannot allow that single incident to destroy the student's performance. You want to devise a system where the student has a chance to understand her or his capability before you allow their enchantment with failure to resurface and they merely stop participating in your course.

The late professor Arthur Bragg (chairman of the mathematics department at Delaware State University) would offer his students an opportunity to drop their lowest examination grade when he calculated their final grade. His philosophy was that everyone has a bad day (A. Bragg, personal communication, 1963). Professor Bragg's marking system kept hope alive because students recognized that one bad examination did not destroy their chances of obtaining a good grade, so most students did not give up in his class.

The Bragg marking system was excellent for the academic mind-set of the 1960s students. Everyone knew that professor Bragg did not compromise on his expectations, so they accepted that he was teaching them perseverance while he elevated their pursuit for excellence. But the Bragg system was not equipped for the advent of the twenty-first century students who were educated in the mind-set of the era of social promotion in public school education.

You may find that many of your background deficient students may have been victims of social promotion. The American Federation of Teachers paint a chilling picture of the fallout of social promotion.

> Today, too many students graduate from high school ill-equipped to do college-level work or perform adequately in entry-level jobs. . . One explanation is 'social promotion'—that is, school systems' practice of moving students from grade to grade regardless of their academic ability to do the work required at the next level.

Social promotion is an insidious practice that hides school failure and creates problems for everybody—for kids, who are deluded into thinking they have learned the skills to be successful or get the message that achievement doesn't count; for teachers who must face students who know that teachers wield no credible authority to demand hard work; for the business community and colleges that must spend millions of dollars on remediation; and for society that must deal with a growing proportion of uneducated citizens, unprepared to contribute productively to the economic and civic life of the nation" (Passing on Failure, 2002).

A holistic look at The American Federation of Teachers' comments suggests that many students may be enchanted with failure, so you need to make sure your grading system does not exacerbate a national crisis. You need to grade in a method that is fair but it makes students toil over a decision to merely give up when perseverance is in order. My experience has shown to make dropping a course difficult for students, only give one test in the first half of that semester: the midterm test. The goal is to make the students toil with a drop decision even if their midterm test grade is bad. The underpinning idea is to have the students invest a significant amount of time (half the semester) in the class concentrating on understanding the material and demonstrating this understanding in blackboard assignments (worth twenty-five percent of grade); to remind students continually that the professor's role is to teach the mathematics and not merely to fail people; and, through use of the Bragg Grading System, to make it very difficult for initially poor-performing students to give up. The hidden agenda here is to teach perseverance. My dropout rates for all courses were very low.

The key idea here is that the students have hope for a better tomorrow. The Bragg Grading System offers that hope, but what impact does it have on top students? Good students may get lazy because they have no need to drop a test to get an "A" for the semester. Therefore, good students may not want to take the final examination or merely show up and put their names on the test paper if attendance is mandatory.

What is very disquieting here is that the class may succumb to the acceptance of mediocrity versus the pursuit of excellence. People may not care whether they fully appreciate the advanced information in the course. Clearly the Bragg Grading System needs some refining to handle the current chasm between offering hope to the majority of students needing background enhancement and excellent students who may become lazy in the advanced sections of the course.

A Modified Bragg Grading System where only one of the first two examinations should be dropped due to a poor grade is encouraging all of the students to pursue excellence in studying the advanced material offered in the latter portion of the course. This grading scheme was proven out in finite mathematics and business calculus courses where class sizes were over thirty students. Under this new Modified Bragg Grading System, no other test grades would be dropped even if they were lower than the first or second score.

Another testing technique is to use an idea that is standard practice in the school of engineering at Cornell University. Engineering students are permitted to bring into the test one sheet of paper with the equations they believe are important to pass the test (S. Miller, personal communication, 1995). This technique focuses the students' attention on understanding the material and not merely memorizing a great deal of equations. It also dissuades students' desire to cheat because they get to bring the equations that they thought were important into the examination.

My experience has shown that giving students a choice of problems to do on the test may reduce their test anxiety. My preference is to give tests with seven problems that are mostly word problems that can have two parts to each problem, so my comments will focus on this design. On test one, students may pick any five of the seven problems offered to complete, but they must circle their selections or I mark the worst ones they tried. On test two, students are required to do the first two problems and then they pick any other three problems from the remaining five problems. On tests three and four students must do the first three problems and pick two remaining problems from the four offered.

Finally, in test two problem numbers one and two touch on subjects that should be comprehended by every student to make the understanding of the advanced material simpler. This grading system will also allow this test to be dropped if it has a lower score than the first examination.

Since the underpinning goal is to have the students learn to dig information out of a textbook to solve real world problems, the third test is an open book examination. The test should be made sufficiently difficult to allow the students to use whatever information they have garnered over the semester. Students may be required to do the first three problems and pick two additional problems from the remaining four problems. This open book technique rewards people who keep good notes. However, a good addendum to this technique is to put all of the key equations needed to solve the problems on the test spread throughout the questions on the test (i.e., the equations in problem seven can be used to solve problem three). If there is a fourth test, it will follow a closed variation of test three.

References

Angel, A., and S. Porter. 1997. *A Survey of Mathematics with Applications*, 5th edition. Reading, MA: Addison-Wesley.

Assessment Report 2000. www.nv.cc.va.us/assessment/AssessReport2000.htm

Aufmann, R. and R. Nation. 1995. *College Algebra and Trigonometry*. New York, NY: Houghton Mifflin.

Barnett, R. and M. Ziegler. 1993. *College Algebra With Trigonometry* 5th edition. New York, NY: McGraw-Hill

Bittinger, M. and M. Keedy. 1995. *Basic Mathematics 7th edition.* Reading, MA: Addison-Wesley

Borich, G. 1996. *Effective Teaching Methods* 3rd edition. Englewood Cliffs, NJ: Prentice Hall.

Brunner, B. 2000–2003. *Timeline of Affirmative Action Milestones.* www.infoplease.com/spot/affirmativetimeline1.html

Classroom Management Profile: Authoritative. 1996. *Teacher Talk, Indiana University, Center for Adolescent Studies.* http://education.indiana.edu/cas/tt/vli2/authoritative.html

College Enrollment, by Sex and Attendance Status: 1983 to 1995. *Statistical Abstract of the United States: 1998.*

Del Junco, T. November 6, 1996. *Statement on Passage of Proposition 209 by UC Board of Regents Chair Tirso del Junco* University of California Board of Regents. www.ucop.edu/ucophome/commserv/press/junco.htm

References

Di Mascio, W. 1997–1999. Seeking Justice: Crime and Punishment in America. New York: *The Edna Mc Connell Clark Foundation*. First Annual-Critical Forum on Prison and Welfare Policies: Economic Exclusion, Public Control, and the Hidden Apartheid. www.public.asu.edu/~ymljus/facts.html Organizing Factsheet United States Student Association www.usstudents.org/foundation/CDP/recretprisons.pdf.

Dolmatch, T.B. 1982. *Information Please Almanac*. 36th edition. New York, NY: Simon & Schuster.

Du Pont, P. 1994.. Education is key to ending welfare. *Journal Gazette*. September 25. Scripps Howard News Service.

Education 173: Learning Theory and Classroom Practice. Department of Education - University of California, course derived from "Effective Teaching, Effective Learning," 2nd edition by Irvine Elliott, S., Kratochwill, T., Kuttkefield, J., Traver J. Benchmark Publishers, 1996 and "Discipline with Dignity" Curwin, R., Mendler A., Association for Supervision and Curriculum Development, Alexandria, VA, 1988. http://olympia.gse.uci.edu/ed173/resources/lectures/unit9_lectures.html

Education Teachers face violence from pupils and parents. February 10, 1999. *BBC Online Network*. Retrieved December 28, 2003 from http://news.bbc.co.uk/hi/english/education/newsid_276000/276356.stm

Engelgau, G. *The Effects of the Hopwood Decision on Texas A & M University's New Freshman Admission*. www.collegeboard.org/aes/ontarget/ontarg12/html/12supp.html.

Erbe, B. 1999. College complaint is mere race baiting. *News Journal*. February 6. Wilmington DE: Scripps Howard News Service

Fact Sheet – Zero Tolerance. *Building Blocks for Youth*. www.buildingblocksforyouth.org/issues/zerotolerance/facts.html

Fishman, B. 1998. "Minority admissions decline at law school." *Daily Bruin*. November 11. Los Angeles, CA: University of California.

Forte, L. 2002, October. *Minority students rate college prep, problems*. www.catalyst-chicago.org/10-02/1002mainprint.htm

General Bill H301: Teachers' Protection Act of 1996. *State of Florida*. www.leg.state.fl.us/session/1996/house/bills/BillInfo/Html/h0301.html

Government releases latest crime statistics. January 13, 1997. *Lubbock Avalanche-Journal*. www.lubbockonline.com/news/011397/governme.htm

Harrison, P. and A. Beck. July, 2003. *Prisoners in 2002*, Bureau of Justice Statistics Bulletin (NCJ 200248), US Department of Justice. www.ojp.usdoj.gov/bjs/pub/pdf/p02.pdf

References

Hedges, M. 1999 Prison populace explodes in U.S. *News Journal.* March 15. Wilmington, DE: Scripps Howard News Service.

Hermanson, K. 1998. Enhancing the Effectiveness of Adult Learning Programs: The Importance of Social and Developmental Learning. *New Horizons for Learning.* www.newhorizons.org/lifelong/workplace/hermanson.htm

Higher Education in Texas: 1998 Status Report. *Blacks and Hispanics Remain Underrepresented*

Johnston, L., R. MacDonald, P. Mason, L. Riley, and C. Webster. 2000. The impact of social exclusion on young people moving into adulthood. *Findings.* www.jrf.org.uk/knowledge/findings/socialpolicy/030.asp

Kataria, P. May 5, 1998. U.S. House debates affirmative action bill. *Pipe Dream Online* - State University of New York (Binghamton) www.pipedreamonline.com/980505/news/n2.shtml

Katz, L. 1967. Eyewitness: The Negro in American History New York, Ny: *Pitman Publishing Corporation*

Kirsch, I., & A. Jungeblut. 1986a. *Literacy: Profiles of America's young adults.* Report No. 16-PL-02, Princeton, NJ: National Assessment of Educational Progress.

Larson, R., R. Hostetler, and B. Edwards.1997. *Algebra and Trigonometry, A Graphing Approach*, 2nd edition. Boston, MA: Houghton Mifflin.

Learning Mathematics At The College Level. www.rit.edu/~369www/MathArt.pdf.

Lewin, T. March 24, 2003. New Online Guides Allow College Students to Grade Their Professors. *The New York Times Archive* March 24, 2003, Late Edition - Final, Section A, Page 11, Column 1, March 24, 2003 http://query.nytimes.com/gst/abstract.html?res=F30711F63F540C778EDDAA0894DB404482&fta=y

Lockman, N. 1995. Who will bring black factions together? *News Journal,* Editorial Page December 26.Wilmington DE

Marable, M. 1999. Education vs. Incarceration. January 27. *Philadelphia New Observer.* Philadelphia, PA

McQueen, A. 1999. Teachers feel unprepared for classrooms *News Journal.* January 29. Wilmington DE: Associated Press.

Miller, S. 1998. *The Quick White Paper*. Wilmington DE: S.N. Miller of Delaware, Ltd.

Miller, S. 1999. The importance of passion in teaching. *Delaware State News* January 25. Dover DE

References

Miller, S. February 15, 1999. Racial and Ethnic Diversity: The Paradigm for the New Millennium Black History Month Keynote Address to Delaware Technical and Community College. http://hometown.aol.com/shermanmil/deltech-speech-21599.htm.

Miller, S. 2003. Should the Federal Government Underwrite the First 60 College Credits? January 16 release to publication at *News Journal*. Wilmington. DE

Miller, S. June 30, 2003. Teaching Today's 100 Level University Mathematics. unpublished manuscript, Delaware State University.

Moffitt, D. 1999. *City Council approves $1,000 scholarship program. News Journal*. February 5. Wilmington, DE

Monticello Research Department. September, 1989. *Education: Jefferson Quotations*. www.monticello.org/reports/quotes/education.html

Nation's Prison and Jail Population Exceeds 2 Million Inmates for First Time. April 6, 2003 US Department of Justice, *Bureau of Justice Statistics*. www.ojp.usdoj.gov/bjs/pub/press/pjim02pr.htm.

Orfield, G., M. Bachmeier, D. James, and T. Eitle. 1997. *Deepening Segregation in American Public Schools* Harvard Project on School Desegregation

Passing on Failure: District Promotion Policies and Practices, PreK–12. 2002. *Educational Issues*. www.aft.org/edissues/socialpromotion/execsumm.htm (no-longer-accessible).

Pick-A-Prof. 2004. www.pickaprof.com/

Quote from Benjamin Franklin. *Liberty-Tree.Ca.* http://quotes.liberty-tree.ca/quotes.nsf/quotes5/60a843a500bc68ad85256cdb001072bd

Reed, R. 1997. *Strategies for Dealing with Troublesome Behaviors in the Classroom*. www.ntlf.com/html/pi/9710/strat.htm

Rigors and Routine. 1998. *The Atlantic Monthly*. www.theatlantic.com/issues/98nov/read3.htm

School Crime 1991. U.S. Department of Justice, *Bureau of Justice Statistics*. www.ojp.usdoj.gov/bjs/abstract/sc.htm

Social Promotion. 2001. *North Central Regional Educational Laboratory*. www.ncrel.org/sdrs/timely/spdef1.htm

Special Analysis 2002 Nontraditional Undergraduates. 2003. *National Center for Education Statistics*. http://nces.ed.gov//programs/coe/2002/analyses/nontraditional/index.asp.

Tall, D. July 13, 1997. From School to University: the Transition from Elementary to Advance Mathematical Thinking. *University of Warwick,*

References

United Kingdom. Presented at the Australasian Bridging Conference in Mathematics at Auckland, University New Zealand.

Taylor, A. 1999. UD's goal applies to diversity. *News Journal*: Wilmington, DE. February 7.

The National Center for Public Policy Research. 2003. Supreme Court of the United States Brown v. Board of Education, 347 U.S. 483 (1954). www.nationalcenter.org/brown.html

Venezky, R., C. Kaestle, and A. Sum. January 1987. The Subtle Danger Reflections on the Literacy Abilities of America's Young Adults, *Center for the Assessment of Educational Progress*, Education Testing Service Report No. 16-CAEP-01.

Wall Street Journal. 1999. Affirmative Action Programs Scaled Back for Highway Projects. February 1.

Wasley, P. 2004. *Responsible Accountability and Teacher Learning* Holding Accountability Accountable. New York NY: Teachers College Press.

Winerip, M. June 6, 2003. A 70 Percent Failure Rate? Try Testing the Testers. *New York Times (nytimes.com)*. www.nytimes.com/2003/06/25/education/25EDUC.html?pagewanted=print&position=

Wright, P.W.D., P. D. Wright, S. Health. 2004. *Wrightslaw No Child Left Behind*. Hartfield, VA: Harbor House Law Press.

Index

150 problems 41, 62

absolute minimum 114
absolute value 116
academic tenacity 43
Added Concerns with Regular Students 45
Addition of a Constant 157
additive inverse 143
affirmative action 6, 13
al Qaeda 16
algebraic mind-set 39
algebraic solution 140
all of your students can learn 167
American Federation of Teachers 171
America's corporate community must have an educate 16
assist sheet 79
asymptote 105
attendance 166

Banking Institution Educational Uplift Program 57
bar chart 96
bell curve 169
bimodal distribution of ages 164
national black leadership pursued the wrong long-t 7
Black Talented Tenth 6
Arthur Bragg 171
Bragg Grading System 172, 172–174, 173
brainteasers 163
busing 17

can teach someone else 49
cartesian coordinates 90
cell addresses 90
center of the circle 95
choice of problems 173
circle 94, 132
circular cylinder 135
city high school student had a 'D' average 18
class participation 166

Index

classroom decorum 167, 170
cleaning the blackboard 167
President Bill Clinton 8
College Try 26
common denominator 38, 59, 87
Community College Role in Reclamation 56
commuters 45
complex numbers 142
compound interest formula 65
conjugate 143
Cornell University 173
course outline 47
creative excuses 45
cube 133
cultural background 170

'D' average 55
deaf ears 70
decision 108
degree of the polynomial is greater than 2 145
Delaware State University 23, 33
Delaware Technical and Community College 56
Sonia Delgado-Tall 19
Descartes' Rule of Signs 151
desegregation remedy 18
develop good work habits 48
difference between mathematics and arithmetic 70
distance formula 92
diversity 13
do not hear excuses 45
domain 83
dos and don'ts 46
drop decision 172
dropped the lowest test score 53
Pierre (Pete) Du Pont IV 15

educational metamorphosis 165
enchantment of failure 50

enchantment with failure 40, 166, 170
equivalent function 81
estimated curve shape 137
Eurocentric immigration 16
even function 115
everyone has a bad day 53
Evolution of a Teaching Tenor 25
exponential instructional model 57

factoring 81
fear of rejection 98
fellows must take off their hats and head rags 167
felony conviction 15, 66
first examination 35
focused on the positive 40
fractions 36
Benjamin Franklin 2
function 98
Function Line Intercepts At One Point 107

Gander Hill Prison 33
Gander Hill Prison Teaching Experience 35
gateway course 161, 167, 168
gauge 98
GED 63
generalized form 80
global economic war 17
Go to the Board Sherman Miller 63
goals and timetables 8
going slow to go fast 41
good grades 164
good working habits 168
graphic solution 135–145

hard science mind-set 163
haughty mannerisms 170
hidden agendas 83, 130, 158
high-energy class 33

Index

high-powered arithmetic 84, 164
homeboys 32
homework assignments 47
'hood 26
Hopwood decision 10

I dont care 141
I get your children and your children's children 51
I need a tissue today 61
imaginary number 142
Imaginary Zeros Theorem 151
importance of blackboard 28
independent axis 82, 90, 116
Indiana University 25
inequality 75
Inmate Course Dynamics — Second Semester 50
Inmate Course Dynamics - First Semester 48
Inmate Pre-calculus Classroom Modus Operandi 46
inmates called upon these pre-calculus students 49
Inner-city America became a wasteland 18
intercepting coordinate values 140
interception points 137–145, 147, 149
intra-classroom racial segregation 23
intuitive 121, 141
inverse function 127
irrational numbers 72

Thomas Jefferson 2
President L. B. Johnson's Executive Order 11246 6
Doretha C. Jordan 31

learning requires stick-to-it-tiveness 130
learning to read a mathematics textbook 49
leg-lengths 93
Less-chanced students 20
lifetime of learning 167
line chart 97
Location Theorem 152
lower bound of the real zeros 151

mainstream receptivity 169
Thurgood Marshall 3
mathematical derivations 79
mathematical modeling 107
mathematical reasoning skill 83
Mathematics Education Research Center 20
mathematics mind-set 106, 163
Mayoral Effort to Encourage Academic Excellence 55
mediocrity 19
memorized 70, 87, 122, 165
memorizing mathematical formulas 122
midpoint 94
Miller Course Level I & III 60
Miller Course Level V 63
minute details 97
Modified Bragg Grading System 173
Debra Moffitt 18
money 156, 164, 165

natural numbers 72
neither even nor odd 116
New teachers 170
President Richard M. Nixon 8
no deviations 106
no eating 167
No one could put himself down 40
no sidebar conversations 167
no working on other course assignments 167
no-excuse mentality 166
Northern Virginia Community College

21
not allow students to put themselves down 166
numerical long division 146

odd function 114–116
older students 163
one to one function 127
one-sided limits 76
one-to-one correspondence 107, 127
one-to-one function 126
operational definitions of racially desegregated 169
opportunity to test some beliefs 31
oral grade 54, 166, 168
Gary Orfield 7
origin 95, 115, 116

parsing definitions 95
pedagogical dispositions 170
PEMDAS 47
perfect square 141
perimeter 131–132
Philadelphia Order 6
physical difference 92
Pick-A-Prof 21
pictorial meaning 70
polynomial of odd degree with real coefficients 151
pragmatic 88
Precalculus Semester I Topic Focus 48
Project Success Wilmington Campus 19
projections 93
proofs 164
Proposition 209 9
Pythagorean Theorem 91–93

racial diversity 170
Radical Equations 154

radicand 83, 103, 142
radius 94, 132, 134, 135
rational numbers 72, 84, 149
reading inertia 130
Real mentorship programs 19
real numbers 72, 74, 81, 83, 89, 100, 101, 102, 106, 111, 112, 125, 126, 127, 128, 142
reasoning ability 92, 135
rectangle 131, 132
rectangular solid 134
Regular Student Course Dynamics — First Semester 51
Regular Student Course Dynamics — Second Semester 53
reputation 86
Responsible Accountability and Teacher Learning 19
U.S. Representative Frank Riggs (R-CA) 11
right triangle 91–93
Rochester Institute of Technology 20
rule 71, 79, 98, 99, 100, 114, 141, 156

Salesianum High School 31
school-busing 4
schools vs. prisons 67
Second Semester 50
self-sufficient adults 165
separate but equal 2
show data in open forums 140
sidewalk philosophers 164
simplistic definition 100
Six-Week Summer Course I 52
Sixteen-Week Course 51
Skill tests 171
slope 101, 108, 109
social promotion 18, 171
socioeconomic parity 7
soul brothers 60
soul brothers and sisters 24

Index

sphere 134
square 91, 131, 132, 141, 155
square root 94, 117, 142
stick-to-it-tiveness 130
student attendance 166
student attrition 166
Student Testing 171
sufficient data 130
Summer Course II 53
syllabus curve 162
symmetric 114–116
synthetic division 148–151

take control 168
take-home tests 46
David Tall 20
teacher intimidation 33
teacher's mind-set 169
Teachers' Protection Act 25
teaching effectiveness 170
Teaching Premises 34
Teaching Through Passion 23
team approach 130
tenacity 130
Testing Initial Conclusions 33
think mathematically 69, 70, 78, 91, 92
Third International Mathematics and Science Study 20
transition to algebraic concepts 37
transitive property 156
trend 97
triangle 93, 133
trifling 62
two million people in jail or prison 66
two-sided limits 76

understanding has an addictive impact 37
understanding is very addictive 26, 49
understanding mind-set 164
uniqueness of numbers 106, 126
units 131–135
University of California 46
University of California Board of Regents 9
University of Delaware 13
unlearn bad habits 167
Upper and Lower Bounds of Real Zeros 151
upper bound of the real zeros 151
use money as the theme 72
user-friendly 106, 141

Richard L. Venezky 4
vertical test 107
vilifying people from crime-ridden neighborhoods 18
violence against teachers 25

Patricia A. Wasley 18, 169
wasted human potential 43
Congresswoman Maxine Waters (D-CA) 11
whole numbers 72, 149
Wolf Ticket 24
word problems 129–134
work at the highest standards possible 167
write down the facts 130

Young Adult Literacy Assessment 4

zero tolerance 25
zeros 140–151

About the Author

Sherman N. Miller is currently a doctoral candidate at the University of Delaware where his area of interest is curriculum and instruction. He is a former visiting instructor of mathematics at Delaware State University and he has been acting director of the Delaware State University Wilmington campus. Miller has taught mathematics and spreadsheets at Delaware Technical and Community College. He has also been a consultant on hose failure analysis after his 1993 retirement from the E. I. Dupont De Nemours & Co. Inc.

Miller's DSU research interest was focused on developing teaching techniques to help inner-city nontraditional students better comprehend mathematics. He developed an unpublished manuscript entitled, "America's Golden Riffraff," that was a first offering of teaching techniques for college mathematics, college algebra, finite mathematics, and business calculus.

Miller earned a Bachelor of Science degree in mathematics at Delaware State University in 1967 and a Masters of Science degree in physics from the University of Delaware in 1972 (thesis title: Photo-stimulated Release of Stored Charge in Multilayer Films of CdS and CdSe).

Miller's non-Dupont work experiences are: Urban Agent at University of Delaware, National Science Foundation Developmental Fellow in the physics department at the University of Delaware,

About the Author

departmental assistant (mathematics) at Delaware State University, production supervisor for the Chrysler Corporation, and management trainee for the International Playtex Corporation.

Miller's political experiences include a run for the office of Lieutentant Governor of the State of Delaware in the 1996 general election and serving as former chairman of the Ethics Commission of the City of Wilmington, Delaware. He has offered marriage seminars where facets of one of the books he and his wife wrote, *Wedlock... The Common Sense Marriage,* are discussed in detail. Miller has given marital advice on national and international radio shows.

Miller continues to write a column on business, politics, and education for newspapers across the United States. His articles have appeared in *Delaware Capitol Review*; the *News Journal* in Wilmington, Delaware; *Delaware State News*; *Philadelphia Daily News*; *Philadelphia Tribune*; *Philadelphia New Observer*; *Kansas State Globe*; *Charlotte Post*; *Baltimore Times*; the *Carolinian*; the *Tri-State Defender*; the *New Pittsburgh Courier*; the *Journal & Guide* (Norfolk VA); the *Connection*; *Frost Illustrated*; and others.

Miller has had letters to the editor appear in *Business Week, Washington Report, Nation's Business, Black Enterprise, Richmond News Leader,* the (Wilmington, DE) *News Journal, Delaware State News,* and *Delaware Medical Journal.*

Miller has had magazine articles appear in the *Capitol Hill, Omega Psi Phi Oracle,* and the Delaware State Chamber of Commerce *Business Journal.*